# Optimizing Microsoft Azure Workloads

Leverage the Well-Architected Framework to boost performance, scalability, and cost efficiency

**Rithin Skaria**

BIRMINGHAM—MUMBAI

# Optimizing Microsoft Azure Workloads

**Group Product Manager**: Preet Ahuja
**Publishing Product Manager**: Preet Ahuja
**Senior Editor**: Shruti Menon
**Technical Editor**: Irfa Ansari
**Copy Editor**: Safis Editing
**Project Coordinator**: Ashwin Kharwa
**Proofreader**: Safis Editing
**Indexer**: Hemangini Bari
**Production Designer**: Ponraj Dhandapani
**Marketing Coordinator**: Rohan Dobhal

First published: July 2023

Production reference: 1110723

Published by Packt Publishing Ltd.
Grosvenor House
11 St Paul's Square
Birmingham
B3 1RB, UK.

ISBN 978-1-83763-292-3

www.packtpub.com

# Foreword

Cloud computing has revolutionized the way enterprises think about the IT infrastructure that enables their business needs. It has provided agility to businesses and enabled them to scale their operations and innovate at an unprecedented pace. However, while cloud computing has lent immense power to the hands of IT departments, it can be a liability if not used responsibly. As organizations move more workloads to the cloud, they face new challenges in managing their cloud costs.

This book is a comprehensive guide to cloud optimization using the Well-Architected Framework. While it is written in the context of Microsoft Azure, the principles articulated can be extended to any hyperscale public cloud. This book provides you with practical advice on how to optimize your cloud environment while maintaining cost, performance, reliability, operational excellence, and security. Rithin has done an excellent job of distilling complex concepts into easy-to-understand language. This book is a must-read for anyone who wants to get the most out of their cloud investment.

Rithin is a great colleague and it was an honor when he reached out to me for an opportunity to preview this book and write this foreword. I am confident that it will be a valuable resource for anyone who is looking to optimize their cloud environment.

*Jatinder Pal Singh*

*Director – Solutions Architecture, Microsoft Qatar*

# Contributors

## About the author

**Rithin Skaria** is a prominent supporter of cloud technologies, in addition to his roles as a speaker, consultant, and published author with a specialization in the design and enhancement of cloud architecture. He has spent over a decade managing, implementing, and designing IT infrastructure solutions for public and private clouds. At present, he works with Microsoft Qatar as a cloud solution architect, placing particular emphasis on Azure solutions. Rithin holds an impressive array of over 18 certifications in diverse technologies such as Azure, Linux, Microsoft 365, and Kubernetes, and he is a Microsoft Certified Trainer. His substantial contributions to the Microsoft worldwide Open Source community have earned him recognition as one of its engagement leaders. He has also spoken at several events and conferences, including Microsoft Spark.

*My heartfelt thanks to my family, manager, colleagues at Microsoft, and everyone who provided their unwavering support and guidance on this journey of writing this book.*

## About the reviewer

**Suraj S. Pujari** is a cloud solution architect at Microsoft India with more than 13 years of experience in IT. His technical capabilities span helping customers with digital transformation, migration, solution designing, and modernizing on the cloud. He works with a wide range of small, medium, and large businesses in the banking and manufacturing domains. In his free time, he likes to do yoga and play with his little one.

*I would like to thank my family and friends for all the support that I have received while staying up late and doing reviews of this book, and finally, the Packt team for giving me this opportunity and bearing with me throughout the process.*

# Table of Contents

# Part 2: Exploring the Well-Architected Framework Pillars and Their Principles

## 3

## Implementing Cost Optimization                                       53

## 4

## Achieving Operational Excellence                                     89

# 5

## Improving Applications with Performance Efficiency                                        115

# 6

## Building Reliable Applications                                                             139

# 7

## Leveraging the Security Pillar                                                             157

# Part 3: Assessment and Recommendations

## 8

### Assessment and Remediation    179

# Preface

This book explores the Microsoft Well-Architected Framework, which is a comprehensive set of best practices and guidelines developed by Microsoft to optimize Azure workloads. By leveraging this framework, cloud architects and developers can build cost-effective, secure, reliable, resilient, and high-performing workloads in Microsoft Azure. The Well-Architected Framework aligns the principles and best practices to five interconnected pillars, which are: Cost Optimization, Operational Excellence, Performance Efficiency, Reliability, and Security. In this book, we will take a consistent approach to evaluate each pillar of the Well-Architected Framework and learn how to optimize and assess Azure workloads. If you are an architect developing solutions in Microsoft Azure, the Well-Architected Framework is something valuable that you should be aware of. Since the Well-Architected Framework is fully managed by Microsoft, all new updates will be constantly amended in the framework so that your architecture reflects the latest practices, architecture patterns, and Azure features.

## Who this book is for

This book is primarily designed for architects and administrators who are already utilizing Microsoft Azure, aiming to improve their understanding and efficiency in optimizing Azure workloads using the Well-Architected Framework.

## What this book covers

*Chapter 1, Planning Workloads with the Well-Architected Framework*, takes you through the concept of the Well-Architected Framework, its pillars, and its elements. The pillars and elements will be explained at a very high level, which will help you to plan your workload optimization.

*Chapter 2, Distinguishing between the Cloud Adoption Framework and Well-Architected Framework*, explores both of these frameworks used in Azure. It's often confusing to decide which framework to use. In this chapter, we will be demystifying the concepts of the Cloud Adoption Framework and the Well-Architected Framework.

*Chapter 3, Implementing Cost Optimization*, explains the importance of cost optimization, cost governance, and relevant tools.

*Chapter 4, Achieving Operational Excellence*, describes all the processes and operations required to keep your mission-critical applications up and running. In this pillar, we deal with the automation of deployment and release processes to eliminate any human error.

*Chapter 5*, *Improving Applications with Performance Efficiency*, explains how to ensure good performance with changing demand. We need to make sure that our workloads don't face any performance degradation during peak hours. With the help of performance efficiency best practices, we can ensure that our workloads can tackle this crisis.

*Chapter 6*, *Building Reliable Applications*, discusses a crucial aspect of maintaining high availability and developing business continuity solutions. In this chapter, we will explore the reliability pillar of the Well-Architected Framework, its patterns, and its best practices.

*Chapter 7*, *Leveraging the Security Pillar*, explores the primary concern for every organization during the deployment of applications in the public cloud, which is the security of these applications. We need to think about security during the complete life cycle of the application as it's an inevitable component. In this chapter, we will see how we can build secure applications.

*Chapter 8*, *Assessment and Remediation*, shows you how to assess your Azure workloads using the Well-Architected Framework and come up with a remediation plan to optimize your Azure environment.

## To get the most out of this book

You should have knowledge about Azure compute, storage, network, governance, identity, data, and monitoring solutions.

| Software/hardware covered in the book | Operating system requirements |
| --- | --- |
| Microsoft Azure | Windows, macOS, or Linux |
| Azure PowerShell | Windows, macOS, or Linux |
| Azure CLI | Windows, macOS, or Linux |

**If you are using the digital version of this book, we advise you to type the code yourself or access the code from the book's GitHub repository (a link is available in the next section). Doing so will help you avoid any potential errors related to the copying and pasting of code.**

## Conventions used

There are a number of text conventions used throughout this book.

`Code in text`: Indicates code words in text, database table names, folder names, filenames, file extensions, pathnames, dummy URLs, user input, and Twitter handles. Here is an example: "A simple search for `3d video rendering` returns two reference architectures."

A block of code is set as follows:

```
for ($i = 1; $i -le 10; $i++){
    Write-Host "Creating ubuntu-vm-$i"
    New-AzVm `
    -ResourceGroupName $resourceGroup`
    -Name "ubuntu-vm-$($i)" `
    -Location 'East US' `
    -Image UbuntuLTS `
    -size Standard_B2s `
    -GenerateSshKey `
    -SshKeyName vmsshkey
}
```

**Bold**: Indicates a new term, an important word, or words that you see onscreen. For instance, words in menus or dialog boxes appear in **bold**. Here is an example: "Once you click on **Start Assessment**, you will be asked to sign in."

> **Tips or important notes**
> Appear like this.

# Get in touch

Feedback from our readers is always welcome.

**General feedback**: If you have questions about any aspect of this book, email us at customercare@packtpub.com and mention the book title in the subject of your message.

**Errata**: Although we have taken every care to ensure the accuracy of our content, mistakes do happen. If you have found a mistake in this book, we would be grateful if you would report this to us. Please visit www.packtpub.com/support/errata and fill in the form.

**Piracy**: If you come across any illegal copies of our works in any form on the internet, we would be grateful if you would provide us with the location address or website name. Please contact us at copyright@packt.com with a link to the material.

**If you are interested in becoming an author**: If there is a topic that you have expertise in and you are interested in either writing or contributing to a book, please visit authors.packtpub.com.

## Share Your Thoughts

Once you've read *Optimizing Microsoft Azure Workloads*, we'd love to hear your thoughts! Scan the QR code below to go straight to the Amazon review page for this book and share your feedback.

https://packt.link/r/1837632928

Your review is important to us and the tech community and will help us make sure we're delivering excellent quality content.

# Download a free PDF copy of this book

Thanks for purchasing this book!

Do you like to read on the go but are unable to carry your print books everywhere?

Is your eBook purchase not compatible with the device of your choice?

Don't worry, now with every Packt book you get a DRM-free PDF version of that book at no cost.

Read anywhere, any place, on any device. Search, copy, and paste code from your favorite technical books directly into your application.

The perks don't stop there, you can get exclusive access to discounts, newsletters, and great free content in your inbox daily.

Follow these simple steps to get the benefits:

1. Scan the QR code or visit the link below:

https://packt.link/free-ebook/9781837632923

2. Submit your proof of purchase
3. That's it! We'll send your free PDF and other benefits to your email directly

# Part 1: Well-Architected Framework Fundamentals

Microsoft has different frameworks designed for every stage of your cloud transformation journey. This section comprises two chapters. In the first chapter, we will go through the fundamentals of the Well-Architected Framework, and in the second chapter, we will cover the differences between the Well-Architected Framework and the Cloud Adoption Framework. Understanding the differences between the frameworks will help you make the right design decisions on your cloud transformation journey. This part contains the following chapters:

- *Chapter 1, Planning Workloads with the Well-Architected Framework*
- *Chapter 2, Distinguishing between the Cloud Adoption Framework and Well-Architected Framework*

# Planning Workloads with the Well-Architected Framework

Microsoft has different frameworks nurtured for Azure; prominent ones are the **Cloud Adoption Framework (CAF)** and the **Well-Architected Framework (WAF)**. There are other frameworks that are subsets of these prominent ones. In this book, we will be covering the WAF and its five pillars.

> **Important note**
>
> Do not get confused with the *Web Application Firewall* in Azure, which is also often denoted as WAF. If you see any reference to WAF in this book, that is the Well-Architected Framework.

Just to give you a quick introduction, the WAF deals with a set of best practices and guidelines developed by Microsoft for optimizing your workloads in Azure. As described in the opening paragraph, this framework has five pillars, and the optimization is aligned with these pillars. Let's not take a deep dive into these pillars at this point; nevertheless, we will certainly cover all aspects of the five pillars as we progress. Further, we will cover the elements of the WAF. When we discuss elements, we will talk about cloud design patterns. This is a lengthy topic, and it's recommended that you refer to the *Cloud Design Patterns* documentation (`https://docs.microsoft.com/en-us/azure/architecture/patterns/`) if you are new to this topic. You will see the pattern names coming up when we discuss design principles, but as patterns are out of scope for this book, we will not take a deep dive into this topic.

In this chapter, we will learn why there is a need for the WAF, its pillars, and its elements.

## Why the WAF?

Microsoft Azure has incredible documentation that can help any beginner to deploy their first workload in Azure. With the help of this well-planned documentation and tutorials, deployment is not a tedious task. Now, the question is: *Are these workloads optimized or running in the best shape?*

When it comes to optimizing, some considerations include the following:

- What is the cost of running this workload?

- What is the **business continuity** (**BC**) and **disaster recovery** (**DR**) strategy?

- Are the workloads secured from common internet attacks?

- Are there any performance issues during peak hours?

These are some common considerations related to optimization. Nonetheless, considerations may vary from workload to workload. We need to understand the best practices and guidelines for each of our workloads, and if it's a complex solution, then finding the best practices for each service can be a weighty task. This is where the Microsoft Azure WAF comes into the picture.

Quoting Microsoft's documentation: "*The Azure Well-Architected Framework is a set of guiding tenets that can be used to improve the quality of a workload.*"

While some organizations have already completed their cloud adoption journey, others are still in the transition and early stages. As the documentation states, this framework is a clear recipe for improving the quality of mission-critical workloads we migrate to the cloud. Incorporating the best practices outlined by Microsoft will produce a high-standard, durable, and cost-effective cloud architecture.

Now that we know the outcome of leveraging the WAF, let's look at its pillars. The framework comprises five interconnected pillars of architectural excellence, as follows:

- Cost optimization

- Operational excellence

- Performance efficiency

- Reliability

- Security

The assessment of the workload will be aligned with these pillars, and the pillars are interconnected. Let's take an example to understand what *interconnected* means.

Consider the case of a web application running on a **virtual machine** (**VM**) scale set. We can improve the performance by enabling autoscaling so that the number of instances is increased automatically whenever there is a performance bottleneck. On the other hand, when we enable autoscaling, we are only using the extra compute power whenever we need it; this way, we only pay for the extra instances at the time of need, not 24x7.

As you can see in this scenario, both performance and cost optimization are achieved by enabling autoscaling. Similarly, we can connect these pillars and improve the quality of the workload. Nonetheless, there will be trade-offs as well—for example, trying to improve reliability will increase the cost; we will discuss this later in this book.

Let's take a closer glimpse of these pillars in the next section.

## What are the pillars of the WAF?

As you read in the previous section, Microsoft has divided its optimization plans, targeting five pillars of architectural excellence. Even though we have dedicated chapters for each of the pillars, for the time being, let's cover some key concepts related to each of the pillars.

The following figure shows the five pillars of the WAF:

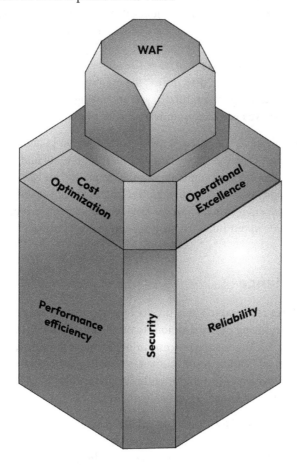

Figure 1.1 – The five pillars of the WAF

We will start with the first pillar, cost optimization.

## Cost optimization

One of the main reasons for organizations to adopt the cloud is its cost-effectiveness. The **total cost of ownership** (**TCO**) is much less in the cloud as the end customer doesn't need to purchase any physical servers or set up data centers. Due to the agility of the cloud, they can deploy, scale, and decommission as required. With the help of the Azure TCO calculator (`https://azure.microsoft.com/en-us/pricing/tco/calculator/`), customers can estimate cost savings before migrating to Azure. Once they are migrated, the journey doesn't end there; migrations mostly go with the lift-and-shift strategy where the workloads are deployed with a similar size as on-premises. The challenge here is that with on-premises, there is no cost for individual VMs or servers as the customer will make a capital investment and purchase the servers. The only cost will be for licensing, maintenance, electricity, cooling, and labor. In the case of Azure, the cost will be pay-as-you-go; for $n$ number of hours, you must pay $n$ times the per-hour cost, and the price of the server varies with size and location. If the servers were wrongly sized on-premises, then during the migration we will replicate that mistake in the cloud. With the servers running underutilized, you are paying extra every hour, every day, and every month. For this reason, we need cost optimization after migration.

It's recommended that organizations conduct cost reviews every quarter to understand anomalies, plan the budget, and forecast usage. With the help of cost optimization, we will find underutilized and idle resources, often referred to as *waste*, and eliminate them. Eliminating this waste will improve the cost profile of your workloads and result in cost savings. In *Chapter 3, Implementing Cost Optimization*, we will assess a demo Azure environment and see how we can develop a remediation plan. Once we figure out the weak points in our infrastructure, we can resize, eliminate, or enforce policies for cost optimization.

## Operational excellence

Operations and procedures required to run a production application are covered by operational excellence. When we are deploying our applications to our resources, we need to make sure that we have a reliable, predictable, and repeatable process for deployment. In Azure, we can automate the deployment process, which will eliminate any human errors. Bug fixes can be easily deployed if we have a fast and reliable deployment. Most importantly, whenever there is an issue post-deployment, we can always roll back to the last known good configuration.

In *Chapter 4, Achieving Operational Excellence*, we will learn about key topics related to operational excellence. For the time being, let's name the topics and explore them later. The key topics are application design, monitoring, app performance management, code deployment, infrastructure provisioning, and testing.

Operational excellence mainly concentrates on DevOps patterns for application deployment and processes related to deployment. This includes guidance on application design and the build process, as well as automating deployments using DevOps principles.

## Performance efficiency

As we saw in the case of cost optimization, we scale the workloads to meet demand with the help of autoscaling; this ability to scale is what we cover in the performance efficiency pillar. In Azure, we can define the minimum number of instances that are adequate to run our application during non-peak hours. During peak hours, we can define an autoscaling policy by which the number of instances can be increased. The increase can be controlled by a metric (CPU, memory, and so on) or a schedule. Nevertheless, we can also define the maximum number of instances to stop scaling after a certain number to control billing. To be honest, this autoscaling scenario was not at all possible before the cloud. Earlier, administrators used to create oversized instances that could handle both peak and non-peak hours. But with Azure, this has changed; the advantage here is that Azure will collect all metrics out of the box, and we can easily figure out bottlenecks.

Proper planning is required to define the scaling requirements. In Azure, how we define scaling varies from resource to resource. Some resource tiers don't offer autoscaling and you must go with manual scaling, while others don't support both automatic and manual scaling. One thing to note here is performance efficiency is not only about autoscaling, but it also includes data performance, content delivery performance, caching, and background jobs. Thus, we can infer that this pillar deals with the overall performance efficiency of our application.

In *Chapter 5, Improving Applications with Performance Efficiency*, we will take a deep dive into performance patterns, practices, and performance checklists.

## Reliability

The word "reliability" means *consistent performance* and, in this context, it means redundant operation of the application. When we build and deploy our applications in Azure, we need to make sure that they are reliable. In our on-premises environment, we use different redundancy techniques to make sure that our application and data are available even if there is a failure. For example, we use **Redundant Array of Independent Disks (RAID)** on-premises, where we replicate the data using multiple disks to increase data reliability.

In Azure or any other cloud, the first and foremost thing we need to admit is that there are chances of failure and it's not completely failproof. Keeping this in mind, we need to design our applications in a reliable manner by making use of different cloud features. Incorporating these techniques in the design will avoid a **single point of failure (SPOF)**.

The level of reliability is often driven by the **service-level agreement (SLA)** required by the application or end users. For example, a single VM with a premium disk offers 99.9% uptime, but if a failure happens on the host server in the Azure data center, your VM will face downtime. Here, we can leverage availability sets or availability zones, which will help you deploy multiple VMs across fault domains/update domains or zones. By doing so, the SLA can be increased to 99.95% for availability sets and 99.99% for availability zones. Always keep in mind that to get this SLA, you need to have at least two VMs deployed across the availability sets or zones. Earlier, we read that the pillars of the

WAF are interconnected, and they work hand in hand. However, in this case, if you want to increase reliability, you need to deploy multiple instances of your application, and what that essentially means is your costs will increase. Remember that these pillars work hand in hand, and sometimes there will be trade-offs, as we have seen in this scenario.

## Security

Security in public clouds was—and is always—a concern for enterprise customers because of the complexity and the way attackers are coming up with new types of attacks. Coping with these types of attacks is always a challenge, and finding the right skills to mitigate these attacks is not easy for organizations. In Azure, we follow the shared responsibility model; the model defines the responsibilities of Microsoft and its customers based on the technology. If we take **Infrastructure-as-a-Service** (**IaaS**) solutions such as VMs, more responsibility is with the customer, and Microsoft is responsible for the security of the underlying infrastructure. The levels of responsibilities will shift more to Microsoft if you choose a **Platform-as-a-Service** (**PaaS**) solution.

It's very important to leverage the different security options provided by Azure to improve the security of our workloads. In the security pillar, we will assess the workloads and make sure they align with the security best practices outlined by Microsoft. As we progress, in *Chapter 7, Leveraging the Security Pillar*, we will take a holistic approach to security and how to build secure applications.

# Exploring the elements of the WAF

Cost optimization, operational excellence, performance efficiency, reliability, and security are the five pillars of the WAF. When it comes to the elements of the WAF, this is different from the pillars. If we place the WAF in the center, then we have six supporting elements. These elements support the pillars with the principles and datasets required for the assessment.

As you know, the WAF is a set of best practices developed by Microsoft; these best practices are further categorized into five interconnected pillars. Now, the question is: *Where exactly are these best practices inscribed?* In other words, the practices should be developed first before we can categorize them into different pillars. This is where the elements come into the picture. The elements act as a stanchion for the pillars.

As per Microsoft's documentation, the supporting elements for the WAF are the following:

- Azure Well-Architected Review
- Azure Advisor
- Documentation
- Partners, support, and service offers
- Reference architecture
- Design principles

Now, we will see the explanation of each of these elements. Let's start with the **Azure Well-Architected Review**.

## Azure Well-Architected Review

Assessment of the workload is required for the creation of the remediation plan; the assessment is inevitable. In the Well-Architected Review, there will be a set of questions prepared by Microsoft to understand the processes and practices in your environment. There will be a separate questionnaire for each pillar of the WAF. For example, the questionnaire for cost optimization will contain questions related to Azure Reserved Instances, tagging, Azure Hybrid Benefit, and so on. Meanwhile, the operational excellence questionnaire will have questions related to DevOps practices and approaches. There will be different possible answers to these questions, varying from recommended methods to non-recommended methods. Customers can answer based on their environment, and the system will generate a plan with recommendations that can be implemented to make their environment aligned with the WAF.

The review can be taken by anyone from the **Microsoft Assessments** portal (`https://docs.microsoft.com/en-us/assessments/?mode=home`). In the portal, you must select **Azure Well-Architected Review**, as shown in the following screenshot:

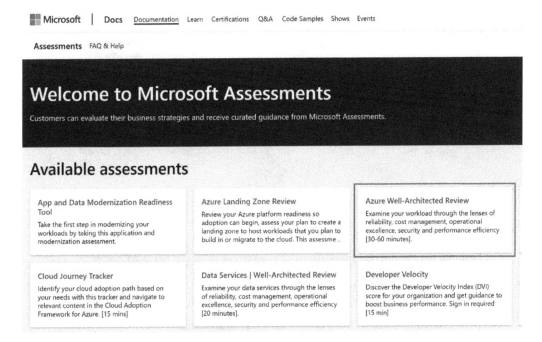

Figure 1.2 – Accessing Microsoft Assessments

Once you select **Azure Well-Architected Review**, you will be presented with a popup asking whether you want to create a new assessment or create a milestone. If you want to create a new assessment, then you can go for **New Assessment**, or choose **Create a milestone** for an existing assessment. At this point, we will conduct an assessment; nevertheless, each pillar of the WAF has its own dedicated chapter, and we will perform the assessment there.

With that, we will move on to the next element of the framework, which is **Azure Advisor**.

## Azure Advisor

If you have worked on Microsoft Azure, you will know that Azure Advisor is the personalized cloud consultant developed by Microsoft for you. Azure Advisor can generate recommendations for you, and you can leverage this tool to improve the quality of workloads. Looking at *Figure 1.3*, we can see that the recommendations are categorized into different groups, and the group names are the same as the pillars of the WAF:

Figure 1.3 – Azure Advisor

With the help of Azure Advisor, you can do the following:

- Get best practices and recommendations aligned to the pillars of the WAF
- Enhance the cost optimization, performance, reliability, and operational excellence of workloads using actionable recommendations, thus improving the quality of the workloads
- Postpone recommendations if you don't want to act immediately

**Advisor** has a score based on the number of actionable recommendations; this score is called **Advisor Score**. If the score is lower than 100%, that means there are recommendations, and we need to remediate them to improve the score. As you can see in *Figure 1.3*, the **Advisor Score** total for the environment is **81%**, and the **Score by category** values are on the right side.

The good thing about Azure Advisor is that recommendations will be generated as soon as you start using the subscription. You don't have to deploy any agents, make any additional configurations, or pay to use the **Advisor** service. The recommendations are generated with the help of **machine learning (ML)** algorithms based on usage, and they will also be refreshed periodically. **Advisor** can be accessed from the Azure portal, and it has a rich REST API if you prefer to retrieve the recommendations programmatically and build your own dashboard.

In the coming chapters, we will be relying a lot on Azure Advisor for collecting recommendations for each of the pillars.

Now that we have covered the second element of the WAF, let's move on to the next one.

## Documentation

Microsoft's documentation has done an excellent job of helping people who are new to Azure. All documentation related to the WAF is documented at `https://docs.microsoft.com/en-us/azure/architecture/framework/`. As a matter of fact, this book is a demystified version of this documentation with additional examples and real-world scenarios.

As with all documentation, the WAF documentation is lengthy and refined, but for a beginner, the amount of information in the documentation can be overwhelming. This book distills the key insights and essentials from the documentation, providing you with everything you need to get started. The following screenshot shows the documentation for the framework:

Figure 1.4 – WAF documentation

As you can see in the preceding screenshot, the contents are organized according to the pillars, and finally, the documentation is concluded with steps to implement the recommendations. You could call this the *Holy Bible* of WAF. Everything related to the WAF is found in this documentation and we would strongly recommend bookmarking the link to stay updated.

All documentation for Azure is available at `https://docs.microsoft.com/en-us/azure/?product=popular`. The documentation covers how to get started, the CAF, and the WAF, and includes learning modules and product manuals for every Azure service. Apart from the documentation, this site offers sample code, tutorials, and more. Regardless of the language you write your code in, Azure documentation provides SDK guides for Python, .NET, JavaScript, Java, and Go. On top of that, documentation is also available for scripting languages such as PowerShell, the Azure CLI, and **infrastructure as code** (**IaC**) solutions such as Bicep, ARM templates, and Terraform.

## Partners, support, and service offers

Deploying complex solutions by adhering to the best practices can be challenging for new customers. This is where we can rely on Microsoft partners. The **Microsoft Partner Network** (**MPN**) is massive, and you can leverage Azure partners for technical assistance and support to empower your organization. You can find Azure partners and Azure Expert **Managed Service Providers** (**MSPs**) at `https://azure.microsoft.com/en-us/partners/`. MSPs can aid with automation, cloud operations, and service optimization. You can also seek assistance for migration, deployment, and consultation. Based on the service you are working with and the region you belong to, you can find a partner with the required skills closer to you.

Once the partner deploys the solution, there will be break-fix issues that you need assistance with. Microsoft Support can help you with any break-fix scenarios. For example, if one of your VMs is unavailable or a storage account is inaccessible, you can open a support request. Billing and subscription support is free of cost and does not require you to purchase any support plans. However, for technical assistance, you need to purchase a support plan. A quick comparison of these plans is shown in the following table:

|  | **Basic** | **Developer** | **Standard** | **ProDirect** |
|---|---|---|---|---|
| Price | Free | $29/month | $100/month | $1,000/month |
| Scope | All Azure customers | Trial and non-production environments | Production workloads | Mission-critical workloads |
| Billing support | Yes | Yes | Yes | Yes |
| Number of support requests | Unlimited | Unlimited | Unlimited | Unlimited |
| Technical support | No | Yes | Yes | Yes |
| 24/7 support | N/A | During business hours via email only | Yes (email/phone) | Yes (email/phone) |

Table 1.1 – Comparison of Azure support plans

A full comparison is available at `https://azure.microsoft.com/en-us/support/plans/`. *Basic* support can only open Severity C cases with Microsoft Support. In order to open Severity B or Severity A cases, you must have a *Standard* or *ProDirect* plan. Severity C has an SLA of 8 business hours and is recommended for issues with minimal business impact, while Severity B is for moderate impact with an SLA of 4 hours. If the case opened is a Severity A case, then the SLA is 1 hour. Severity A is reserved for critical business impact issues where production is down. Having a ProDirect plan offers extra perks to customers, such as training, a dedicated ProDirect manager, and operations support. The ProDirect plan also has a Support API that customers can use to create support cases programmatically. For example, if a VM is down, by combining the power of Azure alerts and action groups, we can make a call to the Support API to create a request automatically.

In addition to these plans, there is a Unified/Premier contract that is above the ProDirect plan and is ideal for customers who want to cover Azure, Microsoft 365, and Dynamics 365. Microsoft support is available in English, Spanish, French, German, Italian, Portuguese, traditional Chinese, Korean, and Japanese to support global customers. Keep in mind that the plans cannot be transferred from one customer to another. Based on your requirement, you can purchase a plan and you will be charged every month.

Service offers deal with different subscription types for customers. There are different types of Azure subscriptions having different billing models. A complete list of available offers is listed at `https://azure.microsoft.com/en-in/support/legal/offer-details/`. When it comes to organizations, the most common options are **Enterprise Agreement (EA)**, **Cloud Solution Provider (CSP)**, and **Pay-As-You-Go**; these are commercial subscriptions. Organizations deploy their workloads in these subscriptions, and they will be charged based on consumption. How they get charged depends solely on the offer type. For example, EA customers make an upfront payment and utilize the credits for Azure; any charges above the credit limit will be invoiced as an overage. Both Pay-As-You-Go and CSP will get monthly invoices. In CSP, an invoice will be generated by the partner; however, in Pay-As-You-Go, the invoice comes directly from Microsoft.

There are other types of subscriptions used for development, testing, and learning purposes, such as Visual Studio subscriptions, Azure Pass, Azure for Students, the Free Trial, and so on. However, these are credit-based subscriptions, and they are not backed up by the SLAs. Hence, these cannot be used for hosting production workloads.

The next element we are going to cover is **reference architecture**.

## Reference architecture

If you know coding, you might have come across a scenario where you are not able to resolve a code error and you find the solution from Stack Overflow or some other forum. Reference architecture serves the same purpose, whereby Microsoft provides guidance on how the architecture should be implemented. With the help of reference architecture, we can design scalable, secure, reliable, and optimized applications by taking a defined methodology.

Reference architecture is part of the application architecture fundamentals. The application architecture fundamentals comprise a series of steps where we will decide on the architecture style, technology, architecture, and—finally—alignment with the WAF. This will be used for developing the architecture, design, and implementation. The following diagram shows the series of steps:

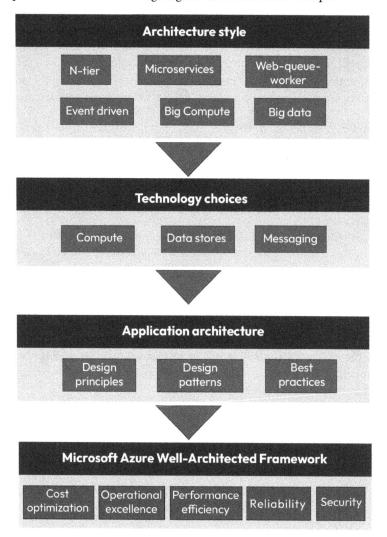

Figure 1.5 – Application architecture fundamentals

In the preceding diagram, you can see that the first choice is the architectural style, and this is the most fundamental thing we must decide on. For example, we could take a three-tier application approach or go for microservices architecture.

Once that's decided, then the next decision is about the services involved. Let's say your application is a three-tier application and has a web frontend. This frontend can be deployed in Azure Virtual Machines, Azure App Service, Azure Container Instances, or even **Azure Kubernetes Service (AKS)**. Similarly, for the data store, we can decide whether we need to go for a relational or non-relational database. Based on your requirements, you can select from a variety of database services offered by Microsoft Azure. Likewise, we can also choose the service that will host the mid-tier.

After selecting the technology, we need to choose the application architecture. This is the stage at which we decide how the architecture is going to be in the following stages and select the style and services we are going to use. Microsoft has several design principles and reference architectures that can be leveraged in this stage. We will cover the design principles in the next section.

The reference architectures can be accessed from `https://docs.microsoft.com/en-us/azure/architecture/browse/?filter=reference-architecture`, and this is a good starting point to begin with the architecture for your solution. You might get an exact match as per your requirement; nevertheless, we can tweak these architectures as required. Since these architectures are developed by Microsoft by keeping the WAF pillars in mind, you can deploy with confidence as these solutions are scalable, secure, and reliable. The following screenshot shows the portal for viewing reference architectures:

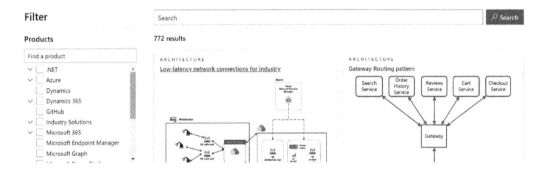

Figure 1.6 – Browsing reference architectures

The portal offers filtering on the type of product and categories. From hundreds of reference diagrams, you can filter and find the one that matches your requirements. For example, a simple search for 3d video rendering returns two reference architectures, as shown in the following screenshot:

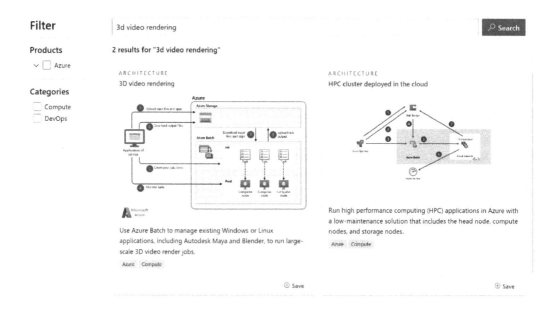

Figure 1.7 – Filtering reference architectures

Clicking on the reference architecture takes you to a complete explanation of the architecture components, data flow, potential use cases, considerations, and best practices aligned with the WAF. The best part is you will have the **Deploy to Azure** button, which lets you directly deploy the solution to Azure. The advantage is the architecture is already aligned with the WAF and you don't have to spend time assessing the solution again.

With that, let's move on to the last element of the WAF—**design principles**.

## Design principles

In *Figure 1.5*, we saw that reference diagrams and design principles are part of the third stage of application architecture fundamentals. In the previous section, we saw how we can use the reference architecture, and now we will see how to leverage the design principles. There are 11 design principles you should incorporate into your design discussions. Let's understand each of the design principles.

## Design for self-healing

As with on-premises, failures can happen in the cloud as well. We need to acknowledge this fact; the cloud is not a silver bullet for all the issues that you faced on-premises but does offer massive advantages compared to on-premises infrastructure. The bottom line is failures can happen, hardware can fail, and network outages can happen. While designing our mission-critical workloads, we need to anticipate this failure and design for healing. We can take a three-branched approach to tackle the failure:

- Track and detect failures

- Respond to failures using monitoring systems

- Log and monitor failures to build insights and telemetry

The way you want to respond to failures will entirely depend on your services and the availability requirements. For example, you have a database and would like to failover to a secondary region during the primary region failover. Setting up this replication will sync your data to a secondary region and failover whenever the primary region fails to serve the application. Keep in mind that replicating data to another region can be more expensive than having a database with a single region.

Regional outages are generally uncommon, but while designing for healing, you should also consider this scenario. Your focus should be on handling hardware failures, network outages, and so on because they are very common and can affect the uptime of your application. There are recommendations provided by Microsoft on how to design for healing—these are called design patterns. The recommended patterns are presented here:

- **Circuit breaker**

- **Bulkhead**

- **Load leveling**

- **Failover**

- **Retry**

As mentioned at the beginning of this chapter, design patterns are not within the scope of this book. Again, thanks to Microsoft, all patterns are listed at `https://docs.microsoft.com/en-us/azure/architecture/patterns/`. Let's move on to the next design principle.

## Make all things redundant

SPOFs in architecture can be eliminated by having redundancy. Earlier, we discussed RAID storage in the *Reliability* subsection of the *What are the pillars of the WAF?* section, where multiple disks are used to improve data redundancy. Azure has different redundancy options based on the service that you are using. Here are some of the recommendations:

- **Understand the business requirements**: Redundancy is directly proportional to complexity and cost, and not every solution requires you to set up redundancy. If your business demands a higher level of redundancy, be prepared for the cost implications and complexity, and the demand should be justifiable. If not, you will end up with a higher cost than you budgeted for.

- **Use a load balancer**: A single VM is a SPOF and is not recommended for hosting mission-critical workloads. Instead, you need to deploy multiple VMs and place them behind a load balancer. On top of that, you can consider deploying the VMs across multiple availability zones for improved SLAs and availability. Once the VMs are behind the load balancer, with the help of health probes we can verify if the VM is available or not before routing the user request to the backend VM.

- **Database replication**: PaaS solutions such as Azure SQL Database and Cosmos DB have out-of-the-box replication within the same region. In addition to that, you can replicate the data to another region with the help of the geo-replication feature. If the primary region goes down, the database can failover to the secondary region for any *read* or *write* requests.

- **Database partitioning**: With the help of database partitioning, we can improve the scalability as well as the availability of the data. If one shard goes down, only a subset of total transactions will be affected; meanwhile, other shards are still reachable.

- **Multi-region deployment**: Regional outages are uncommon; however, we need to account for regional failure as well based on the application requirements. Deploying the infrastructure to multiple regions can help in improving application availability during regional outages. With the help of Azure Traffic Manager and its priority routing, we can failover to the secondary region if the health probe fails.

- **Coordinate failover**: As we discussed in the previous point, we can failover the frontend using Azure Traffic Manager; however, we need to make sure that the database transactions are synchronized to the secondary region and are ready to failover. We need to make sure that when the frontend fails over to the secondary region, the database failover is also coordinated. Depending on the data store that you are using, the failover process may vary.

- **Plan for manual failback**: With the help of Traffic Manager, we can perform automatic failover using health probes, but don't opt for automatic failback. When the primary region recovers from an outage, not all services need to be up and running. For example, let's say the frontend service in the primary region is back online; however, the database is still in recovery. Automatic failback will check if the frontend is up and starts the failback, but the database is not recovered yet. Hence, it's recommended to go with manual failback so that we can verify whether all services are back online and for data consistency to resolve any database conflicts.

- **Plan redundancy for Traffic Manager**: We rely on Azure Traffic Manager for routing traffic in case of regional failure; having said that, the Traffic Manager service can also face downtime. Make sure that you review the SLA of the Traffic Manager service, and if you require more redundancy, consider adding other traffic management solutions as a contingency plan. In case of Traffic Manager failure, we can route the request to the other traffic management solution by repointing our DNS records.

With that, let's learn about the next design principle—minimize coordination.

## Minimize coordination

This principle applies to Storage, SQL Database, and Cosmos DB where we diminish the coordination between application services to accomplish scalability. The key concepts of this design principle are mostly aligned with some data concepts that are not in the scope of this book. The following are recommendations provided by Microsoft for this design principle:

- Consider using the **Compensating Transaction** pattern
- Use domain events to synchronize state
- Use **Command and Query Responsibility Segregation (CQRS)** and **event-sourcing** patterns
- Partition data
- Design idempotent operations
- Consider using async parallel processing
- Use parallel distributed algorithms
- Improve coordination using leader election

An in-depth explanation of these recommendations is available at `https://docs.microsoft.com/en-us/azure/architecture/guide/design-principles/minimize-coordination`.

## Design to scale out

In on-premises, one of the main issues is the capacity constraint. Traditional data centers had capacity issues, and when it comes to the cloud, the advantage is that it offers elastic scaling. In simpler terms, we can provision workloads as required without the need to pre-provision or buy capacity. Talking of scaling, we have two types of scaling, as follows:

- **Vertical scaling**: Changing the CPU, memory, and other specifications of the resource; this is more of a resizing operation. This type of scaling cause the service to reboot. Increasing the size is called scaling up, and reducing the size is called scaling down.
- **Horizontal scaling**: This is where autoscaling comes into context. In horizontal scaling, the number of instances is increased or decreased based on the demand. As there is no change to the initial instance, rebooting is not required, and this process of increasing or decreasing can be automated. Increasing the number of instances is called scaling out, and decreasing the number of instances is called scaling in.

Now that we know the types of scaling, as the name suggests, we need to design for scaling out so that the instances are automatically increased based on the demand. The following recommendations are provided for this design principle:

- **Disable session affinity**: Load balancers have a feature where we can enable session stickiness or session affinity. If we enable this feature, requests from the same client are routed to the same backend server. If there is heavy traffic from a user, the load will not be distributed due to the stickiness, and a single server needs to handle that. Hence, consider avoiding session affinity.

- **Find performance bottlenecks**: Scaling out is not a silver bullet for all performance issues; sometimes, performance bottlenecks are due to the application code itself. Adding more servers won't solve these problems, so you should consider debugging or optimizing the code. Secondly, if there is a database performance issue, adding more frontend servers won't help. You need to troubleshoot the database and understand the issue before choosing to scale out.

- **Identify scaling requirements**: As mentioned in the previous point, different parts or tiers of your application require different scaling requirements. For example, the way the frontend needs to be scaled is not the same way as a database scales. Identify the requirements and set up scaling as required for each application component.

- **Offload heavy tasks**: Consider moving tasks that require a lot of CPU or I/O to background jobs where possible. By doing this, the servers that are taking care of user requests will not be overkilled.

- **Use native scaling features**: Autoscaling is supported by most Azure compute resources. The scaling can be triggered with the help of metrics or based on a schedule. It's recommended that you set up autoscaling using metrics (CPU, memory, network, and so on) if the load is unpredictable. On the other hand, if the load is predictable, you can set up the scaling based on a schedule.

- **Scale aggressively for mission-critical workloads**: Set up autoscaling aggressively for mission-critical workloads as we need to add more instances quickly due to the increased demand. It's recommended that you start the scaling bit earlier than the tipping point to stay ahead of the demand.

- **Design for scaling in**: Just as we scale out, we should design for scaling in. While scaling out, we are increasing the number of instances based on demand; once the demand is gone, we need to deallocate the extra instances that are added during the scaling event. If we don't set up scale-in, the additional instances will keep on running and will incur additional charges.

Now that you are familiar with the scale-out design, let's shift the focus to the next item on the list.

## Partition around limits

In Azure, we have limits for each resource. Some of the limits are hard limits, while others are soft limits. If the limit is a soft limit, we can reach out to Microsoft Support and increase the limit as required. When it comes to scaling, there is also a limit imposed by Microsoft for every resource. If your system is growing tremendously, you will eventually reach the upper limit of the resource. These limits include the number of compute cores, database size, storage throughput, query throughput, network throughput, and so on. In order to efficiently overcome the limits, we need to use partitioning. Earlier, we discussed how we can use data partitioning to improve the scalability and availability of data. Similarly, we can use partitioning to work around resource limits.

There are numerous reasons a system can be partitioned to avoid limits, such as the following:

- To avoid limits on database size, number of concurrent sessions, or data I/O of databases
- To avoid limits on the number of messages or the number of concurrent connections of a storage queue or message bus
- To avoid limits on the number of instances supported on an App Service plan

In the case of databases, we can partition vertically, horizontally, or functionally. Just to give you an idea, let's have a closer look at this:

- In **vertical partitioning**, frequently accessed fields are stored in one partition, while less frequent ones are in a different partition. For example, customer names are stored in one partition that is frequently accessed by the application while their emails are stored in a different partition as they are not frequently accessed.
- **Horizontal scaling** is basically sharding where each partition holds data for a subset of the total data. For example, the names of all cities starting with *A-N* are stored in one partition, while those starting with *O-Z* are stored in another partition.
- As the name suggests, **functional partitioning** is where the data is partitioned based on the context or type of data. For example, one partition stores the **stock-keeping unit** (**SKU**) of the products while the other one stores customer information.

A full list of recommendations is available here: `https://docs.microsoft.com/en-us/azure/architecture/guide/design-principles/partition`. The next design principle we are going to cover is design for operations.

## Design for operations

With the cloud transformation, the regular IT chorus of managing hardware and data center is long gone. The IT is no longer responsible for the data center management as it will be handled by the cloud provider. Having said that, the IT team or the operations team is still responsible for deploying, managing, and administering the resources deployed in the cloud. Some key areas that the operations team should handle include the following:

- **Deployment**: The provisioning of resources is considered deployment, and this is one of the key responsibilities of the operations team. It's recommended that you use an IaC solution for the deployment of services. Using these tools will help reduce human error and makes replication of the environment easy, as templates are reusable and repeatable.

- **Monitoring**: Once the solution is deployed, it's very important that the operations team monitor the solution for any failures, performance bottlenecks, and availability. Having a monitoring system can detect anomalies and notify administrators before they turn into bigger problems. The operations team needs to set up logging and collection to collect logs from all services. The collected logs need to be stored for insights and analysis.

- **Incident response**: As mentioned earlier, we need to acknowledge the fact that failures can happen in the cloud, and if it's a platform issue, the operations team needs to raise a ticket with Microsoft Support. Internally, the operations team can use an **IT service management (ITSM)** solution to create incidents and assign them to different teams for resolution or investigation.

- **Escalation**: If the initial analysis is not yielding any results, there should be processes in place to escalate the issue to the stakeholders and find a resolution. The operations team can have different tiers within the organization that handle different issues; further, they can collaborate with Microsoft Support for issues that require engineering intervention and bug fixes.

- **Security auditing**: Auditing is very important to make sure that the environment is secure. With the help of **security information and event management** (SIEM) solutions, we can collect data from different data sources and analyze them. The operations team can collaborate with external auditors if they lack the necessary skills to perform security auditing. For example, consider using Azure Defender for Cloud and action recommendations. In addition to that, we can use Sentinel to collect data from different sources for analysis and investigation.

A list of recommendations shared by Microsoft can be reviewed at `https://docs.microsoft.com/en-us/azure/architecture/guide/design-principles/design-for-operations`. With that, we will move on to the next design principle.

### Use PaaS services

Unlike on-premises, the cloud offers different service models such as IaaS, PaaS, and **Software-as-a-Service (SaaS)**. Here, we will discuss IaaS and PaaS as SaaS is more of a solution where the end customer doesn't manage the code and is managed by the cloud provider.

In IaaS, the cloud provider takes care of the infrastructure (physical servers, network, storage, hypervisor, and so on) and the customer can create a VM on top of this hardware. Microsoft is not responsible for maintaining the VM OS; it will be the duty of the customer to update, patch, and maintain the OS and code of the application. In contrast, in PaaS, the cloud provider provides a hosting environment where the infrastructure, OS, and framework are managed by Microsoft. The only thing that the customer needs to do is push their code to the PaaS service, and it's up and running. Developers can be more productive and write their code without the need to worry about the underlying hardware or its maintenance.

The design principle recommends using PaaS services instead of IaaS whenever possible. IaaS is only recommended if you require more control over the infrastructure, but if you simply require a reliable environment and ease of management, then PaaS is right for you. *Table 1.2* shows some of the IaaS replacements for popular caches, queues, databases, and web solutions in Azure:

| Instead of running (IaaS) | Consider deploying (PaaS) |
| --- | --- |
| Active Directory | Azure AD |
| RabbitMQ | Azure Service Bus |
| SQL Server | SQL Database |
| Hadoop | Azure HDInsight |
| PostgreSQL/MySQL | Azure Database for PostgreSQL/Azure Database for MySQL |
| IIS/Apache/NGINX | Azure App Service |
| MongoDB/Cassandra/Gremlin | Cosmos DB |
| Redis | Azure Cache for Redis |
| File Share | Azure File Share/Azure NetApp Files |
| Elasticsearch | Azure Cognitive Search |

Table 1.2 – IaaS-to-PaaS considerations

This is not a complete list; there are different ways by which you can replace VMs (IaaS) with platform-managed services. Speaking of services, let's discuss identity services, which are the subject of the next design principle.

### Use a platform-managed identity solution

This is often considered a subsection of the previous design principle; however, there are some additional key points that we need to cover as part of the identity solution. Every cloud application needs to have user identities. Due to this reason, Microsoft recommends using an **Identity-as-a-Service** (**IDaaS**) solution rather than developing your own identity solution. In Azure, we can use Azure AD or Azure AD B2C as an identity solution for managing users, groups, and authentication.

The following recommendations are shared by Microsoft for this design principle:

- If you are planning to use your own identity solution, you must have a database to store the credentials. While storing the credentials, you need to make sure that they are not stored in clear text. In fact, you should consider encrypted data as well. A better option is to perform cryptographic hashing and then salting before persisting the data in the database. The advantage is that even if the database is configured, the data is not easily retrievable. In the past few years, databases storing credentials have been targets for attack, and no matter how strong your hashing algorithm is, maintaining your own database is always a liability. To mitigate this, you can use an IDaaS, where the credential management is done by the provider in a secure manner. In other words, it's the responsibility of the IDaaS provider to maintain the database and secure it. You might be wondering how safe is to outsource the credentials to another provider. The short answer is they have invested time and resources to build the IDaaS platform; if something happens, they are responsible for that.

- Use modern authentication and authorization protocols. When we design applications, use OAuth2, SAML, **OpenID Connect** (**OIDC**), and so on. Don't go for legacy methods, which are prone to attacks such as SQL injection. Modern IDaaS systems such as Azure AD use these modern protocols for authentication and authorization.

- IDaaS offers a plethora of additional security features compared with traditional home-grown identity systems. For example, Azure AD offers passwordless login, **single sign-on** (**SSO**), **multi-factor authentication** (**MFA**), **conditional access** (**CA**), **just-in-time** (**JIT**) access, **privileged identity management** (**PIM**), identity governance, access reviews, and so on. It's going to be a very complex, time-consuming, and resource-consuming task if you are planning to include these features in your own identity system. Above all, the maintenance required for these add-ons is going to be high. If we are using an IDaaS solution, these are provided out of the box.

- The reliability and performance of the identity solution are also a challenge when opting for your own identity solution. What if the infrastructure hosting your identity solution goes down? How much concurrent sign-in and token issuance can happen simultaneously? These questions need to be addressed as they point to the reliability and performance of the identity solution. Azure AD offers SLAs for Basic and Premium tiers, which include both sign-on and token issuance. Microsoft will make sure that uptime is maintained, but in the case of home-grown identity solutions, you must set up redundant infrastructure for keeping the uptime high. Setting up redundant infrastructure is expensive and hard to maintain. Speaking of performance, Azure AD can handle millions of authentication requests without fail. Unlike your own identity solution, IDaaS is designed to withstand enormous volumes of traffic.

- Attacks are evolving and they are getting more sophisticated, so you need to ensure that your identity solution is also evolving and can resist these attacks. Periodic penetration testing, vetting of employees and vendors with access to the system, and tight control need to be implemented. This process is going to be expensive and time-consuming. In the case of Azure AD, Microsoft conducts periodic penetration testing by both internal and external security professionals. These reports are available publicly. If required, you can raise a request for performing penetration testing on your Azure AD tenant.

- Make complete use of features offered by the **identity provider** (**IdP**). These features are designed to protect your identities and applications. Instead of developing your own features, rely on native features, which are easy to set up and configure.

With that, we will discuss the next design principle.

## Use the best data store for your application

Most organizations use relational SQL databases for persisting applications. These databases for good for transactions that contain relational data. Keep the following considerations in mind if your preferred option is a relational database:

- Expensive joins are required for queries
- Data normalization and restructuring are required for schema on write
- Performance can be affected due to lock contention

The recommendation is *not* to use a relational database for every scenario. There are other alternatives, such as the following:

- Key/value stores
- Document databases
- Search engine databases
- Time-series databases
- Column-family databases
- Graph databases

Choose one based on the type of data that your application handles. For example, if your application handles rain-sensor data, which is basically a time series, then you should go for a time-series database rather than using a relational database. Similarly, if you want to have a product catalog for your e-commerce application, each product will have its own specification. The specifications of a smartphone include brand, processor, memory, and storage, while the specifications of a hair dryer are completely different. Here, we need to store the details of each product as a document, and these will be retrieved when the user clicks on the item. For these kinds of scenarios, you should use a document database. In Azure, this type of product catalog can be stored in Azure Cosmos DB.

To conclude, a relational database is not meant for every scenario; consider using alternatives depending on the data that your application wants to store.

We have two more design principles to be covered before we wrap up, so let's move on to the next one.

## Design for evolution

According to Charles Darwin's theory of evolution, species change over time, give rise to new species, and share a common ancestor. The theory also looks at *natural selection*, which causes the population to adapt or get accustomed to the environment. Keeping this theory in mind, when you design applications, design for evolution. This design principle talks about the transformation from a monolithic to a microservices architecture. This transformation is more of an evolution to eliminate tight coupling between application components, which makes the system more inflexible and weaker.

Microservices architecture decouples the application components, and they are loosely coupled. If they are closely packed, the changes in one component will create repercussions in another one. This makes it very difficult to launch new changes into the system. To avoid this, we can consider a microservices architecture, where we can issue changes to the system without affecting other services.

A list of recommendations for this design principle is available at `https://docs.microsoft.com/en-us/azure/architecture/guide/design-principles/design-for-evolution`.

Now, we are going to discuss the last design principle. Let's dive right in!

### Build for business needs

All the principles we discussed so far are driven by a common factor: *business requirements*. For example, when we discussed the *Make all things redundant* design principle, we explored different recommendations for setting up redundant infrastructure. But what if the workload that I have is a **proof of concept** (**POC**) or development workload? Do I need to have redundant VMs for a development workload? As you can imagine, development workloads don't require redundant VMs unless this is demanded by the key factor—*business requirements*. It might seem apparent, but everything boils down to business requirements.

Leverage the following recommendations to build solutions to meet business needs:

- Define business objectives that include certain metrics to reflect the characteristics of your architecture. These numbers include **recovery time objective** (**RTO**), **recovery point objective** (**RPO**), and **maximum tolerable outage** (**MTO**). For instance, a low RTO business requirement needs quick failover automatically to the DR region. On the other hand, you don't have to set up higher redundancy if the business requirement has a higher RTO.

- Define SLAs and **service-level objectives** (**SLOs**) for your application; this will help in choosing the right architecture. For example, if the SLA requirement is 99.9%, we can go for a single VM; however, if the requirement is 99.95%, then you must deploy two VMs in an availability zone.

- Leverage **domain-driven design** (**DDD**), whereby we model the application based on the use cases.

- Differentiate workloads based on the requirements for scalability, availability, data consistency, and DR. This will help you plan the strategy for each workload efficiently.

- Plan for growth; as your business grows, your user base and traffic will grow. You need to make sure your application also evolves to handle the new users and traffic. As we discussed in the *Design for evolution* section, think about decoupling your application components so that your application changes can be easily introduced without disrupting other dependencies.

- On-premises, the cost is paid upfront for hardware, and it's a capital expenditure. With the cloud, on the other hand, there's an operational expenditure, which means you pay for the resources that you consume. Here, we need a shift in mindset because with on-premises, even if you let your VM run for 60 days, there is no additional cost as the hardware cost is paid upfront; the only cost is for electricity and maintenance. But in the cloud, you will be paying for the entire 60 days for which the VM was running. To conclude, delete resources you no longer need to avoid incurring more additional costs than expected.

That was the last design principle, and it's a wrap-up. We have finally completed the elements of the WAF.

## Summary

In this chapter, we started with an introduction to the WAF, and we discussed the five pillars of the WAF. The pillars are cost optimization, operational excellence, performance efficiency, reliability, and security. We briefly covered the concepts and principles of these pillars. Adopting the best practices and recommendations provided by these pillars of the WAF will help you to improve the quality of your Azure workloads.

Then, we discussed the elements of the WAF; recommendations and best practices of the WAF are derived from these elements. In simple words, elements act as the data source for the WAF. There are six elements of the WAF: Azure Well-Architected Review, Azure Advisor, documentation, partners, support, and service offers, reference architecture, and design principles. Understanding these elements will help you learn the best practices that are used to build the WAF. Design patterns and some recommendations for design principles are not included in this chapter as they are out of the scope of the book; nevertheless, you can always refer to the shared links to learn more.

As mentioned in the introduction of this chapter, there are multiple frameworks for the cloud. In the next chapter, we will understand the difference between the CAF and WAF. Readers often tend to get confused between these frameworks, so let's take deep dive into the CAF versus the WAF.

## Further reading

For a deep dive into the topics covered in this chapter, you can refer to the following resources:

- *Microsoft Azure Well-Architected Framework*: `https://learn.microsoft.com/en-us/azure/architecture/framework/`

- *Assess your Architecture with Azure Well-Architected Framework* by *Microsoft Ignite*: `https://www.youtube.com/watch?v=VQ5EfLsjaqA&ab_channel=MicrosoftIgnite`

# 2

# Distinguishing between the Cloud Adoption Framework and Well-Architected Framework

As we saw in the previous chapter, Microsoft is famous for its frameworks and acronyms. The **Cloud Adoption Framework (CAF)** is one of these frameworks and is a close relative of the WAF since they both are designed for Microsoft Azure. We have included this chapter in this book because people often get confused between the Cloud Adoption Framework and the WAF. Although they have different names, the connection between them and the appropriate usage often present a dilemma. Moving forward, we will use the acronym *CAF* for the *Cloud Adoption Framework*, just as we used *WAF* for *Well-Architected Framework*.

In *Chapter 1, Planning Workloads with the Well-Architected Framework*, we introduced the framework and then discussed its pillars and elements. We will take a similar approach for this chapter as well; we will start by introducing the CAF, followed by discussing its methodologies. In addition, we will compare the CAF and WAF. As a side note, the CAF is a very lengthy topic, and we could write a book on each of the methodologies involved; so the introduction and explanation will be limited to a level at which you can distinguish the frameworks. If you are interested in learning the CAF, you can follow the Microsoft documentation (`https://docs.microsoft.com/en-us/azure/cloud-adoption-framework/`).

Compared to *Chapter 1*, this is going to be a shorter chapter as we are only interested in comparing the frameworks. The templates and files described in this chapter are available in Microsoft's official GitHub repository for the Cloud Adoption Framework (`https://github.com/microsoft/CloudAdoptionFramework`). If you are already familiar with the CAF, you can skip to the last section of this chapter, where we will cover the differences between the CAF and WAF.

Let's start this chapter by introducing the CAF.

# Introducing the CAF

Always keep in mind that frameworks are all about best practices, documentation, and tools. Now, which framework to use depends on where you are in your cloud journey. The WAF is used for optimizing workloads that are already in the cloud and to improve the quality of the workload. The CAF, on the other hand, provides end users with best practices, tools, and documentation for adopting the cloud. As its name suggests, the CAF is an adoption framework that's ideal for customers who would like to plan and execute their cloud transformation.

## Purpose of the CAF

In traditional data centers owned by organizations, the organization is responsible for handling the infrastructure, software, and maintenance aspects. The cloud offers different deployment models, such as **Infrastructure-as-a-Service (IaaS)**, **Platform-as-a-Service (PaaS)**, **Software-as-a-Service (SaaS)**, and others, for its end customers to choose from. Depending on their selection and the **shared responsibility model**, part of the responsibility will be handled by Microsoft, while the rest will be handled by the customer. This is one of the most important advantages of the cloud, where the consumers have a lot of flexibility to select the right model. The CAF will act as a guide through your cloud journey.

Though it may look like the CAF is the beginning of the cloud journey, it actually is the last lap. The journey starts when the business stakeholders realize that the cloud can empower and accelerate their business growth. This happens even before they decide which cloud provider to choose. The function of the CAF is to provide the business decision-makers with tools and strategies for the transformation.

The CAF acts as the technical guidance for customers who want to use Azure. Regardless of whether you are going with Microsoft Azure or a multi-cloud strategy, you can still leverage the CAF as it's cloud agnostic when it comes to decision-making. In other words, the strategies outlined in the CAF can be used with any cloud provider.

Let's understand the methodologies of the CAF.

## The life cycle of the CAF

The CAF can be leveraged by architects, administrators, business decision-makers, and other stakeholders for planning their adoption journey. Nine methodologies are listed under the CAF, as follows:

- **Define strategy**: Covers the strategy and justification for choosing the cloud journey. The expected outcomes of the journey are documented.

- **Plan**: The goals behind the reason for choosing the cloud journey are converted into an actionable plan.

- **Ready**: Set up your cloud environment to embrace the changes that are planned.

- **Migrate**: Migrate your existing workloads to the cloud.

- **Innovate**: Reshape workloads to provide more business value by adopting cloud-native or hybrid solutions.

- **Secure**: Enhance the security landscape of your cloud environment.

- **Manage**: Handle cloud and hybrid operations.

- **Govern**: Define organizational standards and add governance to your environment.

- **Organize**: Organize and align teams and roles to handle business areas.

The CAF is an end-to-end life cycle for cloud adoption; it starts with defining the strategy and ends with organizing the resource management teams. The methodologies are a subset of the broader CAF life cycle. The life cycle has different stages to support your journey and overcome roadblocks. The following figure shows the CAF life cycle:

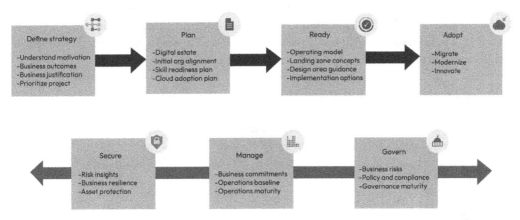

Figure 2.1 – The CAF life cycle

Each life cycle block has processes and methodologies. We will cover these methodologies later in this chapter. For now, let's look at the intended audience for the CAF.

## The intended audience of the CAF

As we mentioned earlier, the CAF is the guidance for businesses to adopt their cloud journey. Here are the intended audiences of the CAF:

- Business leaders

- Business decision-makers

- IT decision-makers

- Finance team
- Enterprise administrators
- IT operations teams
- IT governance
- IT security
- IT policy and compliance
- IT governance
- Solution developers
- Solution owners
- Business SMEs
- Solution SMEs

The guidance provided by the CAF can be leveraged by these roles. In your environment, some of the roles might be combined and handled by a single team. For example, IT governance and IT compliance are closely related and can be handled by a single team. The idea is to include these roles in your cloud journey and make it a success. In the next section, we will take a deep dive into the methodologies of the CAF.

## Exploring CAF methodologies

We saw the nine methodologies of the CAF in the *The life cycle of CAF* section. As discussed earlier, each of these methodologies is quite lengthy as they have their own exercise plans and guidance. Links to the official documentation for every method will be provided as we cover each. Let's start with the first one: defining strategy.

### Defining strategy

Developing a cloud strategy is the first step in the CAF. There could be multiple reasons for an organization to choose cloud computing, some of which are as follows:

- Eliminate or reduce the cost of owning infrastructure
- Empower users to deploy and decommission immediately
- Increase productivity and release time
- Expand business and infrastructure to regions where the organization doesn't have a physical data center
- Scalability and elasticity

These are some of the qualities that on-premises infrastructure couldn't offer, and organizations are empowered to do more with the cloud. To benefit from the cloud, we should start documenting the business strategy so that it can be interpreted by the technical stakeholders and should be acceptable to the business decision-makers.

Strategy development is mostly done for documentation preparation, and Microsoft has provided four steps to document the strategy well. You can reach the desired state of transformation by mapping the business strategies. Let's explore the steps provided by Microsoft.

## Define and document motivations

Motivation should be the answer to the question, why are we moving to the cloud? In other words, what challenges are we trying to address by moving to the cloud? This question should be answered by both technical and non-technical stakeholders. For example, if we ask a technical stakeholder this, we are expecting answers like, "We are moving to the cloud because it's scalable," "We are moving to the cloud because it's agile," and so on. The kind of answers we are expecting from the non-technical crowd include "We would like to reduce the capital investment," and "We would like to bring our product closer to the customer by deploying services across regions without the need to set up physical data centers." These answers are satisfactory as they portray your expected outcomes and motivations. However, if the answer is similar to "We are moving to the cloud because everyone is moving" or "Our CTO asked us to move to the cloud," then it's not easy to derive the outcomes from that. These kinds of answers show that you are uncertain about motivation, and you don't have a clear-cut strategy.

Understanding these motivations is very important as this will be the driving force behind the transformation. These motivations may vary from organization to organization, and it's perfectly alright to have multiple motivations. So long as you have motivation, you can achieve the expected outcome. Microsoft has clearly outlined a set of goals that every organization can evaluate to understand their motivations. Again, this is a joint effort by technical, non-technical, and business stakeholders where each group comprehends their motivations clearly. *Table 2.1* shows the goals that can be achieved by cloud adoption, which organizations can use to determine their motivations:

| Business Events | Migration | Innovation |
| --- | --- | --- |
| Data center exit | Cost savings | Constructing new technical abilities |
| Merger or acquisition | Reducing technical complexity | Scaling to meet demand |
| Reduce capital expenditure | Optimizing IT operations | Scaling to meet multi-region demands |
| Compliance requirements | Improve business agility | Enhance the customer experience |
| Sovereignty requirements | Learn new technical capabilities | Transformation of products |
| Improve IT stability | Scaling to meet demand | Democratization |
| Reduce carbon footprint | Scaling to meet multi-region demands | Sustainability |

Table 2.1 – Business goals

As you can see, these motivations are classified into three major categories. Your motivations might fall into different categories based on your expected outcomes. Chances are that once you start building these motivations, you will see that common motivations are shared between business groups. For example, from a migration standpoint, one of the motivations is *scale to meet demand*; this is also part of innovation. Now, our job is to pick the predominant category with the highest number of motivations. This category will be used for adopting the cloud adoption strategy.

Depending on the predominant category, your strategy development process and priorities will vary. Refer to `https://docs.microsoft.com/en-us/azure/cloud-adoption-framework/strategy/motivations` for more insights.

Now, we will move on to the next step – document business outcomes.

### Document business outcomes

Keeping the business outcome in mind will lead to a successful transformation journey. As you know, cloud adoption is not an easy task; it requires planning and a lot of effort. The process of transformation can be quite time-consuming and will lead to a lot of expenses. The journey is not only the responsibility of the IT team; instead, this is a joint venture by all areas of the business. The involvement and support of all the areas of the business are very important for our success.

Keeping the business outcomes as the cornerstone of the conversation will help to engender transparency and collaboration within the team. We need to collect the business outcomes and document them clearly so that everyone is on the same page. This documentation can be electronically developed or specified via a whiteboard where everyone can contribute and work toward a common goal. Having the outcomes documented will help build outcome-focused discussions.

Microsoft has provided a business outcome template that you can use in the Cloud Adoption Framework GitHub repository (`https://github.com/microsoft/CloudAdoptionFramework`). In the interest of time, we are concluding this section; you can learn more about documenting business outcomes at `https://docs.microsoft.com/en-us/azure/cloud-adoption-framework/strategy/business-outcomes/`.

With that, we will move on to the next step.

### Assess financial considerations

One of the major shifts is going to be the way your finances are handled. In on-premises environments, everything is **capital expenditure** (**CAPEX**), where we make the upfront payment for the hardware or software we would like to purchase. For example, if we want to add a server to our data center, we need to reach out to the hardware vendor, request a quote, pay the cost, and get the server shipped to our on-premises data center. Once the shipment has arrived, we need to complete configuration, patching, licensing, and more. This is a lengthy process and the payment we make is a capital expenditure.

When we move to the cloud, this changes to a pay-as-you-go model. In other words, CAPEX changes to **operational expenditure (OPEX)**. The advantage here is that we don't have to pay anything upfront or invest large capital – we pay for the services that we use based on the number of hours or units consumed. Due to this, organizations can easily expand their business without worrying about financial constraints. In on-premises, if you purchase hardware and it's not getting utilized, the amount you invested is a loss. If you try to resell this hardware, you won't get the amount you invested as it's not a new piece of hardware. But in the cloud, you can decommission infrastructure on the fly, and you don't incur any loss.

Apart from hardware, software, and licensing, organizations can save costs in terms of the IT labor, electricity, and air conditioning required for the data center. Considering sustainability as a goal for your organization, by moving data centers to the cloud, you are reducing your carbon footprint.

Now that you are familiar with the financial considerations, let's understand the technical considerations.

### Interpret technical considerations

We need to admit that the cloud is a new technology and requires additional training. Cloud adoption requires IT professionals and administrators to take on a new learning curve and change their current mindset about infrastructure management. To give some examples, if you want to host a web server on-premises, you are pretty much bound to creating **virtual machines (VMs)** or containers and managing their infrastructure. On the contrary, in the cloud, we can create IaaS solutions, such as VMs or VM scale sets, or PaaS solutions, such as Azure App Service, Azure Container Instances, and Azure Kubernetes Service. We can even create SaaS solutions. With the cloud, we have more options and to adopt these extra options, you need to learn the cloud.

Cloud adoption requires learning and developing skill sets as part of a strategy to maintain and manage the cloud resources once they are migrated. In addition to skill set development, consumers should be aware of the technical benefits of moving to the cloud. Remember, when we started with strategy, we discussed motivations. The technical benefits can be considered as motivations for your cloud adoption journey. The following are some of the technical benefits of adopting the cloud:

- **Scalability**: On-premises data centers have limits; we cannot scale beyond a certain limit, even if there is a legitimate demand for resources for our applications. The cloud offers massive scalability, where we can focus on improving the application delivery by providing resources as per our application requirements without worrying about purchasing additional hardware.

- **Availability**: Availability is directly proportional to cost – if you need more availability, you need to invest more. In on-premises, if you want to implement data center redundancy, then you need to set up a new data center altogether and connect them. However, the cloud has a global presence, and we can set up redundant infrastructure in multiple data centers, zones, or even across regions.

- **Security and compliance**: Protecting your infrastructure against the latest attacks requires more security professionals and skills. In Azure, Microsoft continually performs assessments and testing to make sure the infrastructure is safe and secure. Also, Microsoft Azure has the highest number of industry-standard compliance certifications than all other cloud providers.

Regardless of whether your workloads are long-term or short-term, your aim should be to attain more with your investment. Align the workloads according to your requirements and follow the guidelines of cloud economics. With that, we have completed the first pillar of the CAF. Now, we will proceed to the next stage – creating a plan.

## Creating a cloud adoption plan

In the strategy phase, we outlined the motivations that drive the cloud adoption journey in our environment. In the planning phase, we must transform these goals or motivations into an actionable plan. As we saw earlier, each team will have motivations for cloud adoption, and in the planning process, each team is responsible for converting their respective motivations into an actionable plan.

Four key exercises are outlined in the Microsoft documentation that will help you map your motivations to a plan. The strategy and plan template are available in Microsoft's GitHub repository for the Cloud Adoption Framework: (`https://github.com/MicrosoftDocs/cloud-adoption-framework`). You can use this template document to track the output of each exercise outlined by Microsoft. Let's start with the first exercise – digital estate.

### Digital estate

In digital estate, we conduct **cloud rationalization**. Cloud rationalization is the process of assessing assets and understanding the best path to host them in the cloud. To come up with the best approach, we need to create an inventory of all assets we have in our environment. Once the inventory is ready, we can kick off the cloud rationalization process. While building the rationalization process, the traditional approach is to rely on the Five Rs of rationalization. However, for larger estates, this approach is not efficient and can cause severe confusion while assessing the workloads.

For the time being, let's understand the five Rs of rationalization. For larger estates, you can refer to `https://learn.microsoft.com/en-us/azure/cloud-adoption-framework/digital-estate/rationalize`. As the name suggests, the five Rs are as follows:

- **Rehost**: Rehosting is usually called *lift and shift*, where we make very minimal changes to the architecture of the solution. As the name implies, we are lifting the solution and shifting it to the cloud. An example of this type of migration is migrating on-premises VMs to VMs in the cloud. Rehosting is ideal if you don't have the time and labor to change the architecture.

- **Refactor**: In rehosting, we were migrating VMs as IaaS solutions without changing the overall form factor of the application. However, we are leveraging the PaaS solutions that are available in the cloud and migrating our VMs to PaaS solutions. You are already familiar with the rationale behind the transition from IaaS to PaaS solution. The advantage is that it offers ease of administration and lets developers focus on code development rather than underlying infrastructure management.

- **Rearchitect**: In several cases, you might have a very old application; we usually call this a legacy application. When it comes to the cloud, compatibility is something that we need to confirm before migration. Your application might be incompatible with the cloud due to several reasons – for example, when the design decisions were made, the cloud didn't exist. In these kinds of scenarios, we need to rearchitect the application to make it cloud compatible before the transformation. Another case is where your application is compatible, but not cloud-native. Being cloud-native makes you take advantage of microservices architecture and containerization. By adopting a cloud-native design, you can improve the agility and scalability of the application.

- **Rebuild**: When you calculate the cost and investment in rearchitecting the application as a cloud-native application, if the investment is too hard to justify, you can consider rebuilding the application from scratch. Rebuilding helps you quickly adopt the latest design principles at a cost less than the cost of rearchitecting. Rebuilding accelerates application development, and the final product is cloud-ready.

- **Replace**: There are a lot of SaaS solutions available that could replace your existing applications. If the cost of rebuilding the application is higher than acquiring a license for a SaaS product, then it's best to replace your application with a SaaS solution. For example, let's say you have a homegrown CRM solution that you are planning to rearchitect to make it ready for cloud transformation. When you calculate the investment, the cost is higher than owning a license for a SaaS product such as Dynamics 365. In this case, it's better to replace your solution with the SaaS solution. Apart from the cost calculations, you are already familiar with the advantages that SaaS offers over PaaS and IaaS solutions.

Now, we will transition to the next part of planning, which is alignment.

## Initial alignment

One of the key aspects of cloud transformation is to understand the roles and responsibilities of the people involved in the process. Alignment is required for segregating roles and duties among people, without which we cannot proceed with the transformation.

As its name suggests, this is the initial alignment of people; the full-fledged alignment plan takes time. Most of the time, the initial alignment starts with the list of stakeholders involved in the process and gradually, this gets updated on the fly. As you may have noticed, we do have a dedicated methodology in the CAF called *Organize*, which brings in the true importance of organizational alignment.

When we start the planning process, full alignment is possible as we are in the early stages of cloud adoption. However, you need to make sure that the initial alignment is done and documented. The initial alignment that we propose here will be the baseline for the Organize methodology later. As we reach the Organize methodology, we will fine-tune and tweak the initial alignment for managing the cloud. In this process, we will be mapping people based on their skills; Microsoft provides the following questions that can be asked for mapping purposes. These questions can target a single person or a group of people based on the number of people involved in your cloud adoption journey:

- Who will be taking care of the technical actions or tasks that are part of the cloud adoption plan?

- Who will be held responsible for the team's ability to deliver the technical changes?

- Who will be responsible for implementing the governance controls?

- Who will be held responsible for defining the governance controls?

- Are there any other areas where people need to have responsibility or accountability?

As you might have noticed, the first and third points are clearly for those who implement the actions. On the other hand, the second and fourth points target managers, who are accountable for the team's actions. The last point identifies the team and manager for any other additional capabilities that haven't been aligned yet.

Now that we are familiar with the process of aligning people, let's understand how we can ensure that aligned people have the necessary skills to accomplish the cloud transformation.

### Readiness plan

In the IT industry, the roles and responsibilities of people have been changed or replaced over time. For example, we used to have server administrator roles, and their sole responsibility was to handle the physical servers that are deployed in the organization's data center. With the introduction of virtualization, this role was transitioned to a virtualization administrator role. Cloud transformation not only involves migrating applications from on-premises data centers to the cloud – it involves transforming the entire organization, which includes people. As we embrace cloud adoption, we need to ensure that the people who are aligned with different responsibilities and accountabilities are capable of handling the cloud. We need to make sure that they have adequate skills to manage the workloads in the cloud, similar to how they did on-premises.

People will have anxiety about whether it's hard to learn the cloud or whether they will be replaced with skilled people. This is where having a certain mindset comes into the picture – anxiety can only be eliminated with a growth mindset. People need to come out of their fixed mindset and be ready to learn new things. In short, people should be ready to come out of their comfort zone and explore. Some employees might have 20 or 25 years of experience in the IT industry, but if they approach cloud transformation with a fixed mindset, it can be disastrous. Instead, they should adopt a growth mindset and embrace new changes.

The readiness plan can be expanded into three key areas:

- **Capture concerns**: Team managers should discuss the transition with the team and capture the concerns raised by team members. Apart from mindset issues, one of the common issues is the lack of necessary training. For example, your organization is planning to replace your existing virtual desktop solution with **Azure Virtual Desktop** (**AVD**) and one of the key anxieties will be a lack of knowledge about AVD. Managers should document these concerns so that they can be addressed later as the readiness plan is built.

- **Identify gaps**: There will be certain roles and responsibilities that can be easily trained considering the current knowledge level of the employees. However, there will be certain skills that are not currently available in your organization; this is what we call a gap. If the cost of training the team is too high, you can consider hiring new people. Hiring new people will help you get a person with the right skills. This person can also be leveraged to train other people in your team. The idea here is to map every responsibility and make sure you have people with the right skills to handle that.

- **Cross-team collaboration**: In the cloud, there are hundreds of services and it's not easy for a single person to be the jack of all trades. For example, the team that handles the database should be given training on databases, while the team handling the networking team should be given training for networking. In this way, if a database solution has a VPN connectivity issue, the database and networking team can collaborate and fix the issue. Workflows should be defined for each component so that different departments are aware of whom to reach out to in case of an issue.

So far, we have seen how we can rationalize the digital estate and the initial alignment of people and how the readiness plan should be documented. In the next section, we will consolidate everything that we have collected so far to develop the cloud adoption plan.

### Cloud adoption plan

Planning is required to convert business objectives into substantial business efforts. When building your cloud adoption plan, you can consider the following steps:

- **Validate requirements**: Validate that all the requirements and prerequisites have been completed before creating your plan.

- **Understand your priorities**: Classify and prioritize your workloads. To establish an initial adoption backlog, prioritize your first 10 workloads.

- **Align your resources and assets**: Make sure you have identified resources and assets for the prioritized workloads. These resources or assets can be existing or newly proposed ones.

- **Review rationalization decisions**: Migration or innovation decisions should be validated by reviewing the rationalization decisions.

- **Define iterations and releases**: Make sure the iterations and releases are defined for the adoption. The time blocks specified to do work are called iterations. Any work that needs to be done before making a change to the production process is defined as a release.

- **Define timelines**: At this point, we will not be able to define an exact timeline, but for planning, we need to have a rough estimate of the timelines for documentation purposes.

If you fulfill all the preceding steps, you will have a well-established cloud adoption plan to pursue your journey. Let's move on to the next methodology in the CAF.

## Preparing for cloud adoption

Before initiating the adoption, we need to create a landing zone. A landing zone is applicable for customers who are planning to migrate to the cloud, as well as for customers who plan to start or build in the cloud. In short, regardless of whether migration is taking place or not, we need a landing zone. Preparing for cloud adoption mainly deals with the landing zone's design and implementation. The exercises we'll outline in this section are mandatory for preparing for cloud adoption.

### Cloud operating model

IT operating models used to exist before the existence of cloud technologies and mainly involved business alignment, people alignment, security, governance, compliance, and change management. As we move to the cloud, the operating model also changes – for example, we are taking the hardware assets out of the equation. Hardware assets are replaced by digital assets. Nevertheless, we still require the same set of people and processes. Teams no longer need to focus on the uptime of the physical server as it's the responsibility of the cloud provider. Part of this security will be handled by the cloud provider and part of it will be handled by you, which we call the shared responsibility model. A collection of the preceding processes and procedures is called a **cloud operating model**.

The cloud operating model may vary from organization to organization. Some organizations like to focus more on security controls, some prefer to improve productivity and agility, and others want to go for a set of mixed processes. All this boils down to the nature of the organization and its priorities. During the preparation phase, this will be an opportunity for your organization to define the processes and procedures on how to operate in the cloud.

With that, we will move on to the next topic: Azure landing zones.

### Azure landing zones

Azure landing zones are a vast topic that we could write a book on in terms of their design, implementation, and how they are assessed. In simple terms, an Azure landing zone talks about subscription democratization, where we have multiple subscriptions meant for different types of workloads. Following this architecture will help you build an architecture that is responsible for scalability, security, governance, compliance, networking, and identity.

There are two types of landing zones:

- **Platform landing zones**: A central team for several central teams is split by functions, such as networking, identity, and others. It will deploy subscriptions to deliver unified services. These subscriptions are used for various applications and workloads. Platform landing zones are usually used to consolidate certain essential services for better efficiency and ease of operations. Examples of these essential services include networking components (ExpressRoute, VPNs, firewalls, NVA, Bastion, and so on), identity (domain controllers, Azure Active Directory Domain Controllers, and so on), and management services (Automation Accounts, Log Analytics workspaces, Dashboards, Azure Monitor, and others).

- **Application landing zones**: Unlike platform landing zones, in an application landing zone, we leverage management groups to segregate workloads. Here, we deploy one or more subscriptions for a workload or application. These will be placed under different management groups such as `Online`, `Corp`, `SAP`, and others. These management groups will be placed under a parent management group called `Landing zone`. This hierarchy helps us assign separate policies and access controls for our workloads. Application landing zones have been further subcategorized. Refer to this link to read more: `https://learn.microsoft.com/en-us/azure/cloud-adoption-framework/ready/landing-zone/#platform-vs-application-landing-zones`.

Microsoft has provided a conceptual architecture that organizations can leverage for building their landing zone. Again, this is conceptual and does not apply to all customers. Landing zone implementation can be customized as per your organizational requirements. Refer to `https://learn.microsoft.com/en-us/azure/cloud-adoption-framework/ready/landing-zone/tailoring-alz` to understand how to create landing zones based on your requirements. Microsoft has developed this conceptual architecture based on customer feedback and field experiences, which is available at `https://learn.microsoft.com/en-us/azure/cloud-adoption-framework/ready/landing-zone/#azure-landing-zone-architecture`.

If you feel that the conceptual architecture fits your organizational requirements, then you can use the Azure landing zone accelerator. With the help of well-defined templates from Microsoft, you can create the landing zone's structure from the Azure portal. You can find the landing zone accelerator at `https://learn.microsoft.com/en-us/azure/cloud-adoption-framework/ready/landing-zone/#azure-landing-zone-conceptual-architecture`.

Now, we will move on to the next exercise in preparing for cloud adoption.

### Expedition to the target architecture

The definition of expedition in the Oxford dictionary is a journey started by a group of people with a particular purpose, especially that of exploration, research, or war. Well, cloud adoption is also an expedition that's started by a group of people with a common aim in mind. Interestingly this involves exploration and research but not war. Jokes aside, the common aim here is to reach the target architecture. This expedition is driven by business requirements and the need to innovate. The higher the priority of the business requirements, the faster the journey.

Over time, the organization will gain momentum and iterate the deployed services, processes and procedures, and technical readiness. Having more iteration doesn't mean that the organization will reach the destination quicker; this process takes time. Think of it this way: even if you have the fastest car in the world, it doesn't mean that you can always drive at its top speed. Staying at top speed depends on several parameters, such as the condition of the road, traffic, weather conditions, and more. Similarly, how long the cloud adoption takes depends on several factors, such as organizational size, current technical landscape, and technical skills within the team.

Microsoft's documentation provides an analogy of a trip along a freeway. During the journey, there will be multiple on-ramps that can be used to enter the freeway, but the destination will remain the same. It's a wonderful analogy that talks about the current status of the organization and how they can use an on-ramp to get to the freeway, which will take them to their destination. It's a nice read; you can find it here: `https://learn.microsoft.com/en-us/azure/cloud-adoption-framework/ready/landing-zone/landing-zone-journey#on-ramps`.

Now that we are clear about our journey, let's jump into the last exercise of this methodology.

### Design areas

Not all organizations follow the conceptual architecture for landing zones. The architecture will vary based on the business requirements; that is why we have customized deployment options. Now, the question is how you understand the requirements and align your design areas to them. You need to comprehend the design areas before you choose a deployment option. This exercise should be conducted with utmost care. Each design area is mapped to a methodology in the CAF. If you don't pay attention to these design areas, then it will have an impact on the methodology and have repercussions later.

These design areas can be categorized into environment design areas and compliance design areas. The following table shows the environment design areas, their purposes, and the methodology the area is relevant to:

| Area | Purpose | Relevant Methodology |
|---|---|---|
| Azure billing and Azure AD tenant | For Azure AD tenant creation, enrollment setup, department creation, and account creation | Ready |
| Identity and access management | Required for building the necessary security and aligning the architecture with compliance standards | Ready |
| Network topology and connectivity | Networking, Azure connectivity, and hybrid connectivity are covered | Ready |
| Resource organization | Developing management groups and a subscription hierarchy for segregating workloads and applying them to policy and access controls | Govern |

Table 2.2 – Environment design areas

In terms of the compliance design area, we focus on designing a secure and compliant Azure environment. The rationale behind the compliance design area is to provide the necessary tooling and processes to develop the compliance standards and access controls of your workloads. As you progress on this journey, more iterations and refinements will be made to the policy standards. However, it's important to have the baselines set. The following table shows the compliance design areas, their purposes, and the relevant methodology:

| Area | Purpose | Relevant Methodology |
|---|---|---|
| Security | Implement role-based access controls and security controls | Secure |
| Management | Develop a management baseline for the ongoing operations in the cloud | Manage |
| Governance | Implement policies that can be used to quantify and enforce the governance and compliance standards | Govern |
| Platform automation and DevOps | Automate the deployment of landing zones and the relevant resources with the help of the necessary tooling and templates | Ready |

Table 2.3 – Compliance design areas

To conclude, the aforementioned design areas offer considerations while you're creating a landing zone. Each of these topics is vast and gets quite complex if we don't approach this sequentially during the design process. In reality, these design areas decide the future of your environment and help you make vital choices regarding your environment. So, carefully investigate each of these areas and comprehend what alterations are required in your current design to incorporate these design areas.

With that, we have concluded the *Ready* methodology of the CAF. Now, let's explore the next methodology – adopt.

## Adopting the cloud

The idea behind adopting the cloud is to enhance your organization or business. This is the phase where users start using the cloud and experience its benefits. Adoption is comprised of three main approaches: migration, modernization, and innovation. Each of these approaches has solutions and advantages. Based on the desired goals, the approach that you align to varies. Let's start with the first approach – migrate.

## Migrate

Remember when we discussed the five Rs of rationalization? The first R is *Rehosting*. Though there are numerous ways to migrate your workloads to the cloud, the *migrate* approach focuses on rehosting. As we discussed, rehosting is *lift-and-shift*, where we are not making any changes to our workload. Your on-premises VMs will be migrated as VMs to the cloud. In this approach, the migrated workloads will be deployed as IaaS solutions. Usually, the goals of organizations that take the migrate approach include immediate on-premises data center exit, upscale security, and enhanced operations. The key benefits of taking this approach are cost, security, reliability, and performance. Besides that, organizations don't need to go through the process of acquiring or managing hardware whenever they want to expand their infrastructure.

## Modernize

The *modernize* approach relies on refactoring solutions. When it comes to refactoring, architectural changes will be there, and workloads deployed on-premises will be migrated to the cloud as PaaS solutions. Taking this route will improve efficiency, which means that developers can focus on code development rather than working on the infrastructure, and the overall cost of ownership will be less. Organizations that prefer to go with modernization can achieve goals such as a reduction in technical debt and a chance to modernize their applications and data platforms. In modernization, PaaS solutions are adopted instead of traditional IaaS ones. As you are aware, moving to managed PaaS solutions will lead to a shift in the shared responsibility model; in doing so, infrastructure management becomes the responsibility of the cloud provider. In addition to the benefits that we have seen in the migrate approach, modernization offers business development without them needing to manage the underlying infrastructure. Examples include moving from an SQL DB VM to Azure SQL Database and moving from an unmanaged Kubernetes cluster to a managed Kubernetes cluster such as Azure Kubernetes Service.

## Innovate

The *innovate* approach is aligned with the rearchitect approach, where organizations adopt cloud-native technologies. Choosing cloud-native solutions will help in building customer-centric applications that rapidly scale based on business demands. Goals include repositioning the business, repositioning technical solutions, and more. With this approach, we can improve performance, adaptability, and predictive analytics. By following this approach, we can build predictive solutions by adopting data and applications. Here, the priority is the data that drives all the analytics processes.

Comparatively, the adopt methodology is small, but these approaches are crucial as they decide what features of the cloud you will be experiencing. For example, if you go with the migrate approach, then your solution will be deployed as an IaaS solution where you have to manage the infrastructure, but let's say your goal was to let the cloud provider manage the infrastructure. In this case, instead of migrating, you should have selected the modernize path. Always make sure the approach that you takes aligns with the goals of the organization.

With that, we will move on to the next methodology in the CAF.

## Implementing governance

In a traditional data center, we used to have different compliance standards based on industry standards. Similarly, in the cloud, the scenario is no different – we have to set up governance for enforcing organizational standards for our processes and workloads. When we covered planning, we discussed how planning is an iterative process where we make changes to the initial plan to achieve our business goals. In the same way, cloud governance is also iterative. As the organization evolves, so should its governance. For example, let's say you need to work with a US government agency on a project and the project requires you to comply with the US standards such as FedRAMP. In this case, you need to make the necessary changes to your compliance standards so that everything aligns with the standards demanded by the project. Do not consider governance as a one-time process; instead, periodically verify your compliance scores and make sure you meet all the standards.

You can use the following exercises to set up your initial governance. This needs to evolve as the cloud estate evolves:

- **Draft the baselines**: Start by drafting the baselines for your governance model. This model will help you adopt the cloud. These baselines are decided based on your end goal.

- **Use a benchmarking tool**: With the help of governance, benchmarking tools assess your environment. The assessment should be done for the current state as well as the future state so that you are compliant throughout the adoption process.

- **Establish initial governance baselines**: If you have drafted your baselines, it's time for you to apply or implement these governance standards. You must start with a small set of standards that can improve over time. You will have a minimum viable product by the end of this process.

- **Make improvements**: Since governance is an iterative process, make improvements and make sure all the risks have been addressed with the help of governance controls as you progress.

The governance methodology is a lengthy topic. You can learn more about these exercises at `https://learn.microsoft.com/en-us/azure/cloud-adoption-framework/govern/`.

Now that we've learned about governance, let's look at the next methodology.

## Managing the cloud

Planning is the key ingredient of any cloud adoption. Without strong planning, there will be gaps in the process and you might not be able to achieve your desired goals. In cloud management, as its name suggests, we are planning strategies to manage the workload in the cloud. We need to ensure the reliability and operational excellence of the workloads we are moving to the cloud. Otherwise, if something goes wrong, the uptime of our application will be affected. The following exercises are recommended by Microsoft to ensure that cloud management is done properly for your workloads:

- **Develop business alignment**: Every application will have a business commitment and you need to define cloud management strategies to ensure these commitments are not breached. You can start this process by mapping your workloads to business processes. Once they've been mapped, you need to assign criticality to each of the workloads. Some workloads are mission-critical, while others are not that important. Based on your workloads, rank them based on their criticality. Now that you have a list with the top priority workloads at the top of the list, you need to understand the outcomes of an outage for these workloads. If the impact is high, this means that the investment in cloud management for that workload is high. Alongside this impact, document your commitment. For example, an application impact is much higher if the downtime is greater than 5 minutes. With this documentation, users can understand the top priority workloads, their impact, and the business commitment to end users.

- **Define a baseline**: Azure provides a lot of tools via Azure Monitor to ensure that your application health is properly monitored and the team is notified in case of an outage. By developing a baseline for management, you are defining the criticality classifications, tools, and processes required to attain the commitment offered by the business. You can classify these alerts based on their severity – that is, critical, error, warning, informational, and verbose. With the help of these criticality definitions, teams will be able to respond to events promptly without compromising business commitment. For instance, if the alert is critical, the response time will be much shorter, whereas if the alert is informational, the team will have a higher response time.

- **Expand the baselines**: Every process in the CAF is iterative and there is always room for improvement. Business commitments and needs will change from time to time, so you need to ensure that the baselines are updated and expanded as required. For example, let's say that a less-prioritized workload was later promoted to being a high-priority workload due to business requirements. In this case, the team needs to redefine the baselines that were established earlier for low-priority workloads with the requirements of a high-priority workload.

- **Review advanced operations and design principles**: This is for workloads that have high business commitments. These kinds of workloads demand the need for a richer architectural review. This review is mandatory for finding the single points of failure and establishing countermeasures to deliver business commitments.

You can learn more about these exercises at `https://learn.microsoft.com/en-us/azure/cloud-adoption-framework/manage/`.

Now, let's move on to the next methodology: security.

## Securing the cloud

Security is always a concern for many organizations while working with the public cloud, but the truth is that if you uphold the shared responsibility model and follow the best practices, the cloud is safer than your on-premises environment. As mentioned earlier, cloud adoption is a journey without a fixed final destination. As the journey progresses, your goals and requirements will change, and so will the route you will take. Consider that this is a never-ending evolving process that grows in maturity over time. You might be thinking, if there is no destination, how we are going to achieve the goal? Eventually, you will reach the desired state. The achieved desired state is not static; it will change over time, and you need to make changes to your travel plan accordingly. Through the secure methodology, you learn how to handle security in the cloud.

To be honest, security is a gray area and organizations often find it difficult to secure their environments. With the help of processes, best practices, and security models, the CAF helps you accelerate your security journey. With the lessons and requirements that Microsoft learned from its customers, Microsoft developed this guidance for every organization that wants to move its workloads to the cloud. This guidance also borrowed ideas shared by security organizations such as the **National Institute of Standards and Technology (NIST)**, **Open Group**, and **Center for Internet Security (CIS)**. These organizations are big players in the area of security.

The security principles in the CAF are closely related to other common security concepts, models, and frameworks. Some examples include the Zero Trust Model, the NIST Cybersecurity Framework, and the Open Group core principles white paper.

Besides guidance, the secure methodology also involves mapping roles and responsibilities. We need to follow the principle of least privilege while granting access to resources or environments. This principle articulates the idea of assigning the right set of privileges to identities to complete their day-to-day work; they are neither underprivileged nor highly privileged regarding what they are supposed to do. The transformation that we are bringing in will lead to some changes in the security disciplines. The following are some of the low-hanging fruits that you can easily accomplish. Each of these disciplines is crucial in modeling your security landscape:

- **Access control**: Always apply the principle of least privilege while providing access to your applications, resources, and workloads. This will reduce the impact if an identity is compromised.

- **Monitoring**: The IT team is responsible for monitoring, detecting, responding to, and recovering your environment in case of a security event.

- **Device protection**: Protect all assets and make sure the organization's policies and standards are applied to them.

- **Governance**: There should be documentation on how to handle security events if they occur. Make sure the documentation is updated and followed correctly.

- **Innovation**: If you are using DevOps, introduce DevSecOps to ensure that security is taken care of.

Once the workloads have been migrated, you can use the *Security* pillar in the WAF to assess your workloads and make sure they follow the security best practices.

Now, we will proceed to the last methodology in the CAF – organize.

## Organizational alignment

When we talk about cloud adoption, it's not the workloads that decide the future of your business in the cloud – it's the people. Successful migration is driven by the help of skilled people performing their designated tasks according to the organizational requirements. Organizational alignment is a topic we discussed earlier in the planning phase, where we came up with the initial alignment. The following exercises are recommended by Microsoft to help you with the process of creating a landing zone for cloud adoption:

- **Define the structure**: We need to define the roles and responsibilities of teams so that they are aware of their responsibilities. The structure we define here might not be the same as the organizational hierarchy. For example, there will be infrastructure architects reporting to one manager and data architects reporting to another manager. When we look at the organizational chart, they report to different managers. However, when there is a migration that involves infrastructure and data components, a structure will be defined for the collaboration of these architects, even though they report to different managers.

- **Cloud functions**: Several functions have been defined so that you can align your staff based on their maturity. The following are some of the cloud functions that you can implement in your organization to align staff. To some extent, all these functions are used in all cloud adoptions. Sometimes, the same team will handle two functions, while in others, there will be dedicated teams:

  - **Cloud strategy**: For developing strategies for converting technical change into business needs
  - **Cloud adoption**: For technical solution delivery
  - **Cloud governance**: For risk management
  - **Central IT team**: For IT operation management
  - **Cloud operations**: For cloud operations management
  - **Cloud platform**: For managing the cloud platform
  - **Cloud automation**: For automating solutions and accelerating adoption
  - **Cloud data**: For handling data and analytics solutions
  - **Cloud security**: For handling cloud security

- **Mature the team's structure**: Like all the methodologies we've discussed so far, organizational alignment should be an iterative process. The access and function of team members should be reviewed periodically, and amendments should be made accordingly.

- **RACI metrics**: RACI metrics is an Excel template that is used to map roles. Using this will help your teams understand their roles and responsibilities in a better way. Let's expand the acronym:

- **Responsible**: The person who is responsible for completing the task
- **Accountable**: The person who is making the decisions or actions on the task
- **Consulted**: The person who will be consulted throughout the decision process
- **Informed**: The person who will be kept posted on the actions

With that, we have covered all the methodologies in the CAF. Some of the methodologies were covered in depth due to their compatibility with the WAF, while others were briefly covered. Nevertheless, here are some references for all the methodologies we've discussed in case you want to take a deep dive:

| Methodology | Reference |
|---|---|
| Strategy | `https://learn.microsoft.com/en-us/azure/cloud-adoption-framework/strategy/` |
| Plan | `https://learn.microsoft.com/en-us/azure/cloud-adoption-framework/plan/` |
| Ready | `https://learn.microsoft.com/en-us/azure/cloud-adoption-framework/ready/` |
| Adopt | `https://learn.microsoft.com/en-us/azure/cloud-adoption-framework/adopt/` |
| Govern | `https://learn.microsoft.com/en-us/azure/cloud-adoption-framework/govern/` |
| Manage | `https://learn.microsoft.com/en-us/azure/cloud-adoption-framework/manage/` |
| Secure | `https://learn.microsoft.com/en-us/azure/cloud-adoption-framework/secure/` |
| Organize | `https://learn.microsoft.com/en-us/azure/cloud-adoption-framework/organize/` |

Table 2.4 – References for the CAF methodologies

We covered these topics to help you understand what the CAF is before we compare it to the WAF. Now, we will get to the core of this chapter, which is comparing the CAF and WAF.

## Comparing the CAF and WAF

At this point, you know that the CAF and WAF are close relatives and that they share a lot of common ideologies. To recap, the WAF has five pillars: Cost Optimization, Operational Excellence, Reliability, Performance Efficiency, and Security. On the other hand, the CAF has eight methodologies: Strategy, Plan, Ready, Adopt, Govern, Manage, Secure, and Organize.

The CAF is a full end-to-end framework that helps organizations get started with the cloud and migrate their workloads. As its name suggests, it is an adoption framework. For a successful migration, we need strategy, planning, readiness, and governance. These methodologies are covered in the CAF. The CAF helps customers transition from a zero cloud to a full-fledged cloud transformation. This is the starting point for an organization to engage with Azure and gradually migrate to the cloud. The methodologies are arranged in a systematic orderly fashion that can guide the customer. Each methodology has a set of exercises that will empower the customer to achieve more as they progress with their cloud adoption. The CAF not only provides these theoretical concepts but also a set of tools such as Cloud Adoption Strategy Evaluator and the Governance Benchmark tool for organizations to quickly assess their environments and understand where they stand. As discussed earlier, the CAF is a journey without a static destination. Once we migrate the workloads to the cloud, they need to be optimized and managed based on organizational requirements; this is where the WAF comes into the picture.

The WAF targets workloads that are migrated by following the CAF's methodologies. We can leverage the WAF to optimize these workloads based on various aspects, such as cost, operational excellence, performance, reliability, and security. In short, the WAF is only applied when your workload is in the cloud; the WAF cannot be used to assess the workloads that are currently deployed in on-premises data centers. If you would like to assess a workload using the WAF that is on-premises, then you have to migrate the workload to the cloud using the CAF and then assess it using the WAF.

In short, the WAF targets a workload that is already in the cloud; its optimization can be aligned with any of the WAF pillars, and the CAF is used to engage and empower customers who want to migrate their workloads to the cloud. If you take a closer look, you will understand that they work intimately with customer workloads – one for migration and another to optimize it.

I hope this provides you with some clarity and helps you differentiate between the frameworks.

## Summary

In this chapter, we covered the basics of the CAF so that we can finally distinguish between the CAF and WAF. We started this chapter by introducing the CAF and discussing how we can leverage the CAF to engage customers and accomplish cloud transformation. The CAF starts with strategy, where we discuss and align the business priorities. Then, we saw how we can make the organization ready for its cloud journey. Once readiness has been ensured, we proceed with the preparation phase and cover the cloud operating model and landing zones. After this, the next process in the framework is adoption, where we decide what adoption strategy we want to take. The next steps in the process involve governance, cloud management, security, and organizational alignment of the cloud.

Toward the end of this chapter, we covered how the WAF is different from the CAF. Just to recap, in short, the CAF is for new customers who would like to migrate their workloads to the cloud, while the WAF targets workloads that are migrated to the cloud. With that, we concluded this chapter. In the next chapter, we will move on to the first pillar of the WAF: Cost Optimization.

# Part 2: Exploring the Well-Architected Framework Pillars and Their Principles

The Well-Architected Framework is a set of best practices and recommendations developed by Microsoft to optimize your cloud environment. The recommendations and optimizations are aligned to five pillars, namely, Cost Optimization, Operational Excellence, Performance Efficiency, Reliability (formerly known as High Availability), and Security. In this section, we will cover each of these pillars and their respective principles, design areas, recommendations, and best practices. The following chapters are part of this section:

- *Chapter 3, Implementing Cost Optimization*
- *Chapter 4, Achieving Operational Excellence*
- *Chapter 5, Improving Applications with Performance Efficiency*
- *Chapter 6, Building Reliable Applications*
- *Chapter 7, Leveraging the Security Pillar*

# 3

# Implementing Cost Optimization

Organizations migrate to the cloud for various reasons, such as scalability, elasticity, agility, global presence, and so on. In addition to these reasons, there is one more factor that accelerates the rate of cloud adoption, and that is **cost savings**. Business stakeholders are more likely to be interested in understanding the potential **return on investment (ROI)** rather than the technical jargon associated with the technology. From the business point of view, they need to lower the capital expenditure and **total cost of ownership (TCO)**. Organizations perform TCO and ROI analysis before they embrace the cloud journey.

We know that in the cloud, we have the *pay-as-you-go* model. By adopting this model, organizations do not need to invest money in buying or procuring hardware; they will be charged based on consumption. If there is a need to scale the environment, we can leverage the agility and scalability of the cloud without needing to buy hardware. We can easily increase or decrease the number of instances and pay only for what we use, thanks to **autoscaling** features in the cloud. In the **Cloud Adoption Framework (CAF)** in the previous chapter, we learned that the adoption of the cloud is a continuous journey, and stakeholders need to optimize the environment based on business requirements. After the adoption of the cloud is complete, we can't just rely on the principles of the CAF; instead, we need to make use of the **Well-Architected Framework (WAF)** to optimize our workloads.

The first pillar that we have in the WAF is **cost optimization**. In this chapter, you will learn the fundamentals, techniques, principles, and assessment ideas for cost optimization in your Azure environment. Let's get started with an introduction to this concept.

## Introducing cost optimization

As we just discussed, business stakeholders are very much concerned about the ROI that they get from cloud adoption. Although the cloud is not a new concept, we lack people with the right skills to manage the cloud. Due to this very same reason, a misconfiguration could lead to major cost implications.

Nevertheless, Azure provides tools and solutions to ensure that there is predictivity and tracking of costs. The cost optimization pillar offers a set of principles to ensure that your cloud workloads are optimized and that the cost aligns with the business goals and requirements without diminishing the ROI. In short, the idea is to eliminate waste, eradicate unnecessary expenses, and improve cost efficiency without compromising performance.

In on-premises, we procure hardware based on the business forecast. Buying hardware is a capital expenditure, and if the business does not go as expected, the purchased hardware is a liability. When it comes to the cloud, we don't need to pre-provision the servers; the best idea is to use the pay-as-you-go approach and add autoscaling rules. We will be keeping a bare minimum of instances to run your application; these instances will be enough to cover the normal usage of the application. Let's say there is a sudden surge in the number of users; the attached auto-scaling policy will increase the number of instances to cover the unexpected spike. Once the surge is settled, the number of instances will be lowered back to the bare minimum number of instances that we deployed at the beginning. This is how we should design workloads in the cloud to optimize the cloud cost.

## Tools for optimizing cost

With the help of tools, we can estimate the initial cost and cost of maintaining workloads in the cloud for a prolonged period. If that calculation turns out to be positive with higher ROI, then we need to come up with policies, controls, and measures to ensure that ROI stays the same.

In Azure, we have a TCO calculator (`https://azure.microsoft.com/en-us/pricing/tco/calculator/`), which we use to estimate savings by providing the details on your on-premises infrastructure and compare the cost of owning the same in Microsoft Azure. To do this, we need to perform the following steps:

1.  We need to define our workloads as the number of servers, capacity of servers, databases, storage, and networking in our on-premises infrastructure, as shown in the following screenshot:

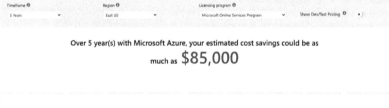

**Define your workloads**

Enter the details of your on-premises workloads. This information will be used to understand your current TCO and recommended services in Azure.

**Servers**

Enter the details of your on-premises server infrastructure. After adding a workload, select the workload type and enter the remaining details.

Figure 3.1 – Defining workloads in TCO

2. The next step is to adjust assumptions where we add the details of licensing, storage, electricity, IT labor, and so on.

3. Finally, once you submit the assumptions, the TCO calculator will produce the savings:

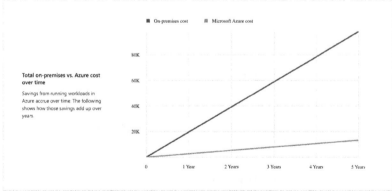

Figure 3.2 – Evaluating estimated savings

> **Important note**
> Though this value doesn't play any role in the WAF cost optimization, this evaluation should be done before migrating to the cloud. This will help in understanding the savings and can be set as a business goal to optimize the cloud once the adoption is fully complete.

The next tool to introduce is the **Azure Pricing Calculator** (`https://azure.microsoft.com/en-us/pricing/calculator/`). With the help of the Pricing Calculator, we can estimate the operational cost of a service in Azure. This will help us in predicting the cost, setting the budget, and forecasting the cost. In the following screenshot, we can see how the cost of running a D2v3 Windows **virtual machine** (**VM**) in the West US region is estimated using the Azure Pricing Calculator:

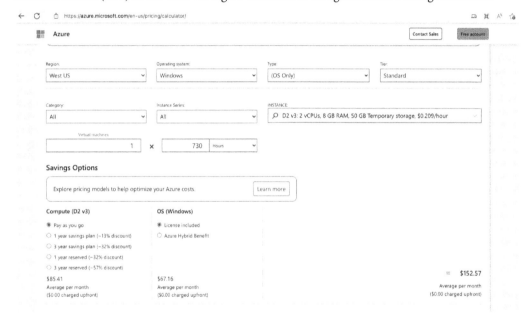

Figure 3.3 – Estimating costs for Azure services

By combining these two tools, we can estimate the ROI and operational cost in Azure. Now we have a basic understanding of cost optimization and the related tools. Let us shift our focus to the next topic.

## Understanding the cost optimization design principles

If a workload runs within a defined budget and is driven by ROI, we can call it a cost-effective workload. **Cost optimization principles** are a set of considerations defined by Microsoft that can improve the overall cost profile of your environment. In addition to that, these principles can assist you in achieving both business objectives and cost justification. Let us understand these principles in detail in the following sub-sections.

# Choosing the right resources

Capturing the exact requirement is the key ingredient in cost optimization. If we capture the right requirements, then we can forecast the budgets correctly, and later, there won't be any problems. In on-premises, if you want to host a web server, the process is simple. Either you deploy the server on a VM or on a physical server. However, in the case of the cloud, we have multiple options, such as the following:

- Deploy the web server on a VM

- Use Azure App Service to host the website

- Use containers in Azure Container Instances

- Use **Azure Kubernetes Services (AKS)**

There are multiple possibilities in the cloud; you can choose IaaS, PaaS or even SaaS solutions to host your application. If you choose PaaS solutions instead of IaaS solutions, the management overhead will be eliminated, as Microsoft is responsible for managing the underlying infrastructure. Also, PaaS solutions are cheaper compared to IaaS solutions.

A misconfiguration or wrong selection can lead to cost implications. For example, let's assume your application uses Azure Storage and the expected **service level agreement (SLA)** to process requests and reading data is 99.9%. In this case, we are good with **locally redundant storage (LRS)** as it offers 99.9% for read operations; instead, if you go for a **geo-redundant storage (GRS)** account, which provides 99.99%, the offered SLA is more than what we need, and the cost will be on the higher side. The bottom line is to understand the requirements and choose the right resources that meet the business needs. Also, we need to ensure that we are not overdoing the cost optimization and degrading the performance of the workload.

The next principle is setting up budgets.

# Setting a budget and cost constraints

With the help of **budgets**, we can bring **accountability** to your environment. This is like having a credit limit for your credit card so that you do not overspend. Every service, architecture, or solution we use comes with a price and each one of them has its own cost implications. Ideally, the company should add the cost constraints before choosing the following:

- An architectural pattern such as sidecar, retry, and so on

- Azure service

- The pricing model of the service

Having predefined cost constraints will help the architects to develop the solution with the overall cost in mind (architecture patterns, Azure service, and the pricing model) and stay within the cost constraint enforced by the organization.

During the design phase, we need to recognize the acceptable boundaries on the following:

- Scale

- Redundancy

- Performance

The more scalability, redundancy, and performance that you need, the more the cost that you pay. The right approach is to find a balance between these and identify acceptable boundaries. For example, due to cost constraints, we might not be able to deploy our solution across multiple regions.

Once the initial cost for running the infrastructure is calculated, then we need to set up budgets and alerts at different scopes. With the help of budgets, we can set up alerts at multiple threshold levels to make the team aware of the cost and act accordingly.

## Allocating resources dynamically

In **dynamic resource allocation**, we need to increase or decrease the number of instances, shut down underutilized instances, and resize resources based on the business requirements. Let us take an example for each of these scenarios and understand how we can optimize cost.

### Scenario 1

**A workload that needs to increase or decrease the number of instances based on users using the application**

In on-premises, we maintain a fixed number of instances to host the application, and this approach may lead to higher costs in the cloud. We need to use autoscaling in this case, where we need to configure the minimum and maximum number of instances and also the metrics for scaling. The following screenshot shows the usage of an application:

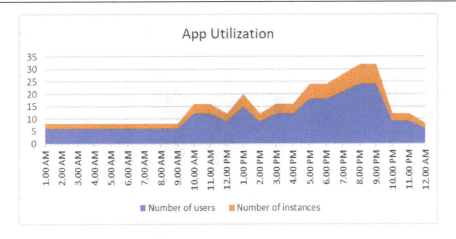

Figure 3.4 – Application utilization

In this case, you can see that if the number of users increases, the number of instances needs to increase proportionally. We could increase or decrease the number of instances manually, but that is inefficient, and we need someone to monitor this workload 24 x 7. With autoscaling, we can define the minimum instances we need, and if the average CPU utilization of the workload goes above, say, 75%, we can increase the instances. This will be our **scale-out rule**. Similarly, we can decrease the number of instances if the average CPU utilization of the instances is less than, say, 25% and eventually roll back to the minimum instances we configured. This will save a lot of costs as you are not running a fixed number of instances and will only pay for the instances based on the hours they consume. At the same time, you are not compromising the performance of your application by increasing the instances to meet the demand. If there is more demand, scaling out happens, and when the demand goes down, we **scale in**.

### Scenario 2

**A workload (single instance) that is consumed heavily during weekdays and consumes less than 10% during the weekend**

In autoscaling, we increase or decrease the number of instances, and these instances have the same configuration running the same application. But if your workload is running on a single instance that is consumed heavily during weekdays and less during weekends, here we cannot take the autoscale approach. The following figure shows the weekly average consumption of the workload:

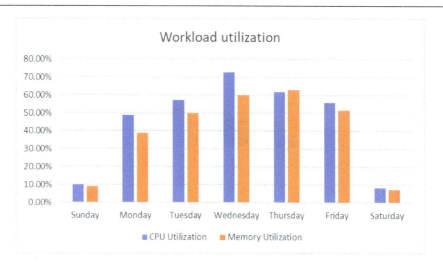

Figure 3.5 – Workload utilization

Here we can create automation to resize the VM during weekends. Let us say we need a single Linux VM *Standard_D16_v3* (16 vCPU, 64 GB RAM), which is ideal for our weekday usage, and when we finish Friday business hours, we need to resize this to a *Standard_D4_v3* (4 vCPU, 16 GB RAM) to accommodate the weekend usage. Let us compare the pricing for this configuration with a per-hour cost of $0.77 in the East US. The following table shows the cost without resizing the VM:

| VM Size | Day | Time | Per-day cost |
|---------|-----|------|--------------|
| Standard_D16_v3 | Sunday | 12:00 A.M. – 12:00 A.M. | $ 18.43 |
| Standard_D16_v3 | Monday | 12:00 A.M. – 12:00 A.M. | $ 18.43 |
| Standard_D16_v3 | Tuesday | 12:00 A.M. – 12:00 A.M. | $ 18.43 |
| Standard_D16_v3 | Wednesday | 12:00 A.M. – 12:00 A.M. | $ 18.43 |
| Standard_D16_v3 | Thursday | 12:00 A.M. – 12:00 A.M. | $ 18.43 |
| Standard_D16_v3 | Friday | 12:00 A.M. – 12:00 A.M. | $ 18.43 |
| Standard_D16_v3 | Saturday | 12:00 A.M. – 12:00 A.M. | $ 18.43 |
| | | **Total** | $ 129.02 |

Table 3.1 – Cost of VM without resizing

The total cost incurred for running the compute resource is $129.02. We have not included the storage and network cost here. Now, we will implement resizing using a *Standard_D4_v3*, which is charged at $0.19/hr from Friday 5:00 P.M. to Monday 9:00 AM. The following table shows the cost with resizing:

| VM Size | Day | Time | Per-day cost |
|---------|-----|------|--------------|
| Standard_D4_v3 | Sunday | 12:00 A.M. – 12:00 A.M. | $ 4.61 |
| Standard_D4_v3 | Monday | 12:00 A.M. – 9:00 A.M. | $ 1.71 |
| Standard_D16_v3 | Monday | 9:00 A.M. – 12:00 A.M. | $ 11.52 |
| Standard_D16_v3 | Tuesday | 12:00 A.M. – 12:00 A.M. | $ 18.43 |
| Standard_D16_v3 | Wednesday | 12:00 A.M. – 12:00 A.M. | $ 18.43 |
| Standard_D16_v3 | Thursday | 12:00 A.M. – 12:00 A.M. | $ 18.43 |
| Standard_D16_v3 | Friday | 12:00 A.M. – 05:00 P.M. | $ 13.06 |
| Standard_D4_v3 | Friday | 05:00 P.M. – 12:00 A.M. | $ 1.33 |
| Standard_D4_v3 | Saturday | 12:00 A.M. – 12:00 A.M. | $ 4.56 |
| | | **Total** | $ 92.08 |

Table 3.2 – Cost of VM with resizing

With resizing, the total cost came down to $92.08, giving us savings of $36.94 per week. If we consider approximately 52 weeks in a year, then the yearly savings is $1,920.88. We need to assess our workloads periodically and understand their utilization, and if you see low utilization on weekends, then this approach will help you optimize the cost.

## Scenario 3

**A workload (single instance) that is consumed heavily during weekdays and is not being used during weekends**

This scenario usually comes up where there are development workstations deployed in the cloud that developers use on weekdays. When it is the weekend, these machines are not required. As we saw in Scenario 2, we can use automation to optimize the cost. Instead of resizing, we are now going to shut down the VM. For example, let us take the *Standard_D16_v3* as our VM size; this means that if we run the VM 24 x 7; the cost will be $129.02 per week (refer to *Table 3.1*). With the help of automation, we will shut down the VMs by Friday at 5:00 P.M. and bring the VM back online on Monday at 9:00 AM when business hours start. The following table shows the cost per day:

| VM Size | Day | Time | Per-day cost |
|---------|-----|------|--------------|
| Standard_D16_v3 | Sunday | 12:00 A.M. – 12:00 A.M. | $       - |
| Standard_D16_v3 | Monday | 12:00 A.M. – 9:00 A.M. | $       - |
| Standard_D16_v3 | Monday | 9:00 A.M. – 12:00 A.M. | $    11.52 |
| Standard_D16_v3 | Tuesday | 12:00 A.M. – 12:00 A.M. | $    18.43 |
| Standard_D16_v3 | Wednesday | 12:00 A.M. – 12:00 A.M. | $    18.43 |
| Standard_D16_v3 | Thursday | 12:00 A.M. – 12:00 A.M. | $    18.43 |
| Standard_D16_v3 | Friday | 12:00 A.M. – 05:00 P.M. | $    13.06 |
| Standard_D16_v3 | Friday | 05:00 P.M. – 12:00 A.M. | $       - |
| Standard_D16_v3 | Saturday | 12:00 A.M. – 12:00 A.M. | $       - |
| **Total** | | | $    79.87 |

Table 3.3 – Cost of VM with shutdown

To conclude, with the ability to control the resources, we can optimize the cost. You need to find which scenario suits your workload and perform the respective optimization. With that, we are moving on to the next principle.

## Aiming for scalable costs

As we have seen in the previous principle, the beauty of the cloud is its agility and ability for dynamic scaling. The cost of the workload should be directly proportional to the demand. If the demand increases, the cost increases, which is ideal. However, if there is no demand and you are paying a cost equal to high demand does not make sense. This is where the need to leverage autoscaling comes in, and the following are some recommendations if you want to aim for scalable costs:

- To determine the number of instances, we need to review the **historic usage metrics** of the service. Let us say we have a VM, and we take the average CPU and memory utilization for the past three months. Assume that the value for CPU and memory utilization is 8% and 7%, respectively. In this case, we can understand that we need to scale down the instance to a smaller one as the full potential of the VM is not being used. Similarly, on the other hand, if you see high utilization, it means we need to add more instances to accommodate the load.

- We can also use autoscaling if we see variable utilization and performance bottlenecks, or if there is more demand, we will add more instances and reduce them automatically when there is no need with the help of autoscaling rules. The right approach is to use smaller instances and add more if there is demand. If we use a larger instance with autoscaling, the newly added instances will be expensive.

- Rather than taking a fixed-size approach by scaling up and down, we need to use **scale-in** and **scale-out** approach. This will help you to optimize the cost in a better way.

In scalable costs, cost management should have the following characteristics:

- **Regularly assessing operational cost**: Each service in Azure has its own meters, which Microsoft charges you for. If you have a considerable number of resources, cost management can become complicated and may appear like a tedious task unless you assess it thoroughly and carefully. Most customers deploy resources in the cloud and forget about it; they never assess the operational cost unless and until they see a spike in the cost. Make sure you perform precise cost management to keep the cost under control.

- **Adopting iterative cost management**: In the CAF, we saw that many processes are iterative; similarly, cost management is also an iterative process. Best practice is to evaluate your cost every quarter, compare it with the previous quarter's cost, and forecast the cost for the next quarter. Adopting iterative cost management will help you to review the cost frequently and adjust the resources or forecast as per business requirements.

- **Ensuring accountability**: By setting up a budget, we can bring in accountability. With the help of budgets, we can notify stakeholders if their cost exceeds the set threshold. Best practice is to set multiple notification thresholds. For example, set notifications at say 60%, 70%, 80%, and 90% of the budget. By calculating the duration between these alerts, we can understand how quickly our budget is getting consumed.

With that, we will move on to the last principle in cost optimization.

## Continuous monitoring and optimization

As we have seen, scalable costs, cost management, and optimization are iterative processes. They do not end in one or two iterations; we need to keep monitoring the cost to ensure that there is continuous optimization. In order to scale with demand and dynamically provision resources, we need to consider the following principles:

- **Treat cost management as a periodic process and conduct regular assessments of your environment**: Business requirements change, and we need to optimize the cost according to that. For example, you deployed a *Standard_D16s_v3* VM to host your website, and the VM was getting fully utilized. After a couple of months, you noticed that the number of requests coming to the website is less compared to previous months. If you don't have a periodic cost review, chances are there that you miss the resizing opportunity. Now that the business doesn't require a *Standard_D16s_v3*, we can resize this to, say, a *Standard_D4s_v3* to optimize the cost.

- **Always follow the T-shirt size approach**: If you walk into an apparel store to buy a T-shirt, you will start with a size closer to your requirements, say large. Let us say the large size is not the right fit, then you will try the extra-large size, and if that also doesn't fit, then you will go to the extra-extra-large size. We need to take a similar approach in the cloud as well, we need to start with the smallest size that meets your requirements. If you are experiencing performance bottlenecks with the selected size, then you will upsize to the next size and keep doing this exercise until you reach the right size. You might wonder why we cannot start with the largest size available and downsize. Though it may look acceptable initially if we deploy with the largest size and keep it running for a week to understand if that is the right size, that will incur higher costs than starting from a lower size and upsizing. To optimize the workload, it is always better to start lower and scale up based on metrics.

- **Consider the forecasting of our capacity needs**: To explain this point, let us take an example of an application that is used for tax filing purposes. You are planning to deploy the application at the beginning of the financial year; you select the VM size to host the application, calculate the estimated cost using Azure Pricing Calculator, and send the estimate to the finance team for approving the budget. So far, it's good, and you are on the right track, but have you accounted for or forecasted the capacity needed for the last days of tax filing? On the last day, there will be so many users trying to file their taxes, and the initial setup will not suffice. Now at this point, we cannot run back to Finance to approve more budget for the workload. Even though you can approve it, you need to provide justification for why there is a spike in the cost and why this was not forecasted. As part of continuous monitoring and optimization, you need to measure the current capacity needs and forecast future capacity needs.

With that, we have concluded the design principles for cost optimization. The points that we discussed are generic and can be applied to any cloud provider's cost optimization. Now we will look at the design checklist.

## Cost design checklist

As the name suggests, a design checklist contains a list of items that you need to consider while designing a cost-effective workload. The checklist has been divided into a **cost model checklist** and an **architecture checklist**. The cost model emphasizes the nitty-gritty of the cost features, while the architecture checklist talks about specific architectural decisions that could impact the overall cost of the solution. Although these are two separate lists, they work together to optimize the cost. Let us start with discussing the cost model checklist.

### Cost model checklist

The cost model deals with the pre-deployment considerations for cost. For example, you plan to deploy a solution in Azure, and before jumping into architecture, you need to understand the economies of scale. Let's discuss the considerations that contribute to the cost model:

- **Capture and comprehend the requirements**: This is the starting point where you need to plan for your requirements. Basically, we would like to capture the aim behind moving to the cloud. There could be multiple reasons, such as advanced features in the cloud, scalability, agility, decommission on-premises environment, and so on. Understanding the requirement is important because we need to ensure that the requirement is fulfilled after migrating to the cloud.

- **Estimate initial cost**: As we learned at the beginning of this chapter, we need to leverage tools such as the Azure Pricing Calculator and TCO Calculator to estimate the cost. The TCO Calculator will be used during the migration to estimate the savings from moving to the cloud. Azure Pricing Calculator can be used to estimate the cost of running services in Azure. We need to produce a **proof-of-concept** (**PoC**) deployment, which will contain all the services that you will be using in production. At this stage, you will have a better understanding of the cost required for production. Also, we need to ensure we choose the right resources while estimating the cost. For example, if we take a smaller VM, the cost might be lower, but when you are dealing with production, this may affect performance. The only option is to go for a larger VM, which will cost you more than the estimate. Furthermore, you need to account for **networking costs** and **replication costs**. Based on the type of replication that you are selecting for services such as storage, there will be replication costs. Similarly, if you download a large amount of data, this will lead to bandwidth costs.

- **Define cost constraints**: We need to define boundaries and budgets for our consumption. With the help of budgets in **Azure Cost Management**, we need to come up with a budget for each department or team. We need to discuss the budget constraints, notification preferences, and what to do when we are over the budgeted value. If we cross the budget value, what is the maximum acceptable value?

- **Shared resources**: Every organization will have a set of resources that will be shared across multiple business units or departments. For example, if you are deploying the Azure landing zone, we will have a hub subscription where we will keep all the shared services that will be consumed by all other business units. It makes sense to deploy services such as ExpressRoute, Firewall, Backup, Log Analytics, and Azure Bastion in a centralized subscription; deploying these services on individual subscriptions can increase the cost drastically, and it will be management overhead. Now that we have resources that are shared by multiple teams, we need to build a charge-back strategy so that we can charge the business units for their consumption. In some organizations, IT is responsible for the cost of shared services, while in others, they equally divide the cost between business units that are using the shared services.

- **Governance**: With the help of Azure Policy, we can control the cost. As you know, Azure resource tags can be sent to Azure Cost Management for cost analysis. Most of the time, users will skip this part and deploy the resources, thus leaving them unaccounted for. Using Azure Policy, we can enforce a set of tags as per your organizational standards. Without filling in values for these tags, users cannot deploy the resource. Similarly, users might deploy expensive VM sizes even if they don't need those. With the help of policies, we can enforce a list of sizes that are allowed to deploy. Furthermore, if you want to have stringent policies, you can define the set of resources users can deploy in their subscriptions.

With that, we have completed the checklist in the cost model. Now let us discuss the architecture checklist.

## Architecture checklist

The architecture checklist consists of a set of considerations that you need to include while designing cost-effective solutions in Microsoft Azure. The following are the key points that you need to consider:

- **Azure regions**: Microsoft Azure has a global presence, so without the need to have a physical data center, we can deploy our workloads to any of the Azure regions. Most people do not know that in Azure, each region has different pricing. For example, if you estimate the cost for the same VM size in US East and Central India, US East will be cheaper. On the other hand, we cannot deploy everything in US East if your users are based out of India. This will lead to latency and performance issues. While choosing Azure regions, there are certain factors you need to consider:

  - **Compliance**: Your organization might have data residency and data sovereignty requirements where they cannot store the data outside their home country. For example, if your organization follows **General Data Protection Regulation** (**GDPR**), then it cannot store data outside the European Union. So, we need to make sure that we are not breaking any compliance rules while running behind the cheapest region.

  - **Performance**: As mentioned earlier, if we compare US East and Central India, US East is cheaper. However, it does not make sense to deploy our resources in US East if our users are in India, just for the sake of cost. If the users do not like your application due to latency and higher load time, then your business will not boom as you forecasted. You can consider a set of regions that offer lower latency, such as Central India, South India, and West India, and pick the cheapest of these. So, it is better to consider regions with lower latency rather than sabotaging your business.

  - **Cost**: Though we are discussing cost optimization, we cannot exclude compliance and performance from the equation to save cost. The correct approach would be to shortlist a set of regions, verify compliance, verify performance metrics, and choose the cheapest on the list.

  - Secondly, opt for geo-redundant architecture only where it is required. For example, if your application has an SLA of 99.95%, which can be achieved by deploying to availability zones, then there is no need to go for geo-redundant architecture. Obviously, this may offer a higher SLA than availability zones; however, that is not needed for your application and would require a higher cost.

- **Subscription type**: There are distinct types of subscriptions that you sign up for in Azure. Microsoft offers Dev/Test subscription, which is ideal for development and testing purposes, as the name implies. The advantage of these subscriptions is you can have lower prices for VMs and other eligible services. One tradeoff here is there is no SLA offered by Microsoft for the services deployed in Dev/Test subscriptions. This makes complete sense, as we are not expecting to meet any SLA for our development workloads. So, with this tradeoff, we can enjoy lower pricing for development and testing workloads. The following subscriptions can be used for development and testing:

- If you already have an Enterprise Agreement with Microsoft, you can sign up for **Enterprise Azure (EA) Dev/Test subscription** (`https://azure.microsoft.com/en-us/offers/ms-azr-0148p/`) to avail of the benefits. You can convert your existing subscription to Dev/Test by reaching out to Microsoft Support.

- If you don't have an EA but have a pay-as-you-go subscription, then you can go for **Pay-As-You-Go Dev/Test** (`https://azure.microsoft.com/en-us/offers/ms-azr-0023p/`).

- If you are a beginner and would like to learn Azure, then there is the free trial (`https://azure.microsoft.com/en-us/free/`), which provides $200 credit for 30 days and other free services.

- If you already have a Visual Studio Enterprise/Professional subscription, then you can sign up for an **MSDN subscription**, which offers credits that will be renewed every month (`https://azure.microsoft.com/en-us/pricing/member-offers/credit-for-visual-studio-subscribers/`).

By using these non-production subscriptions, we can save costs. If you are hosting production resources, make sure you use a production subscription so that SLA covers you. There are other subscription types available for Azure, and you can review all Azure offers at `https://azure.microsoft.com/en-us/support/legal/offer-details/`. Make sure you review the document and pick the right subscription that matches your requirements.

- **The right resources**: As mentioned earlier in this chapter, each Azure service has multiple usage meters, and you need to be aware of this. For beginners, you can refer to the Azure Pricing Calculator. In the pricing calculator, you can review different meters available for a service and estimate the cost. To give an example, if we need the cost for Azure Storage, the initial assumption would be since it is storage, there will be a cost for the amount of data stored. Well, that is correct, you will be charged for the data stored but along with other meters such as transaction cost, retrieval cost, redundancy cost, bandwidth, and so on, as shown in the following screenshot:

Figure 3.6 – Storage cost meters

Now that we know the cost meters, we also need to ensure that the workload is capable of managing the performance. During the initial design, we might pick a cheaper VM; however, over time we need to resize the VM to accommodate the business needs and performance requirements. As this is an iterative process, review and optimization should happen periodically. Finally, if you are using any third-party services from Azure Marketplace, you should be clear about the pricing. Some offerings in Marketplace expect you to bring your own license to use the service, while others will add a charge to your Azure cost. This cost is in addition to your Azure infrastructure costs. The following screenshot shows the cost of Red Hat Enterprise Linux 9.1 offered by AskforCloud LLC, as you can see there is a cost of *$0.038 per hour (or equivalent based on your billing currency) + Azure infrastructure costs* for using this offering:

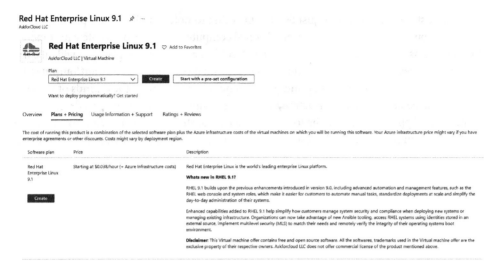

Figure 3.7 – Marketplace cost

Be aware of the cost before you deploy any Marketplace item, navigate to the **Plans + Pricing** section to understand the pricing terms, as shown in *Figure 3.7*.

- **Pre-provisioned versus consumption-based models**: In Azure, we have certain services such as Azure Functions, Azure SQL Database, and so on, which offer a pre-provisioned model as well as a serverless model. The serverless model is also known as the consumption-based model, where you only pay for the time you consume the service. In the pre-provisioned model, the infrastructure is provisioned during the deployment and you are charged at a fixed rate regardless of the consumption. For example, if we take Azure Function; it offers three models: Consumption (serverless), App Service plan (pre-provisioned), and Premium plan (pre-provisioned). In the consumption model of Azure Functions, you will be charged for the number of executions. The following screenshot shows the cost of Azure Functions in the consumption model with 1,500,000 executions (1,000,000 are free of cost):

Figure 3.8 – Consumption model cost

Here the cost for the function is just $0.20. One thing to note here is consumption runs on a shared compute, and there are no pre-provisioned compute resources to execute your function; due to this reason, you might face a **cold start**. A cold start is where the platform will take time to allocate the resources before executing a function, and if your function is **input/output (I/O)-** intensive, this could lead to a delay and performance degradation. In these kinds of scenarios where you need dedicated infrastructure to quickly execute the function, you need to opt for the Premium or App Service plan. The following screenshot shows the cost of running a function in the Premium plan with the lowest configuration possible at the time of writing this book:

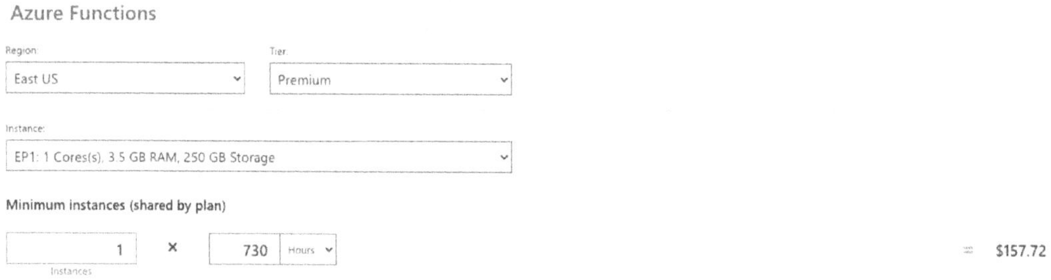

Azure Functions

Region:
East US

Tier:
Premium

Instance:
EP1: 1 Cores(s), 3.5 GB RAM, 250 GB Storage

Minimum instances (shared by plan)

1   ×   730   Hours   =   $157.72
Instances

Figure 3.9 – Premium plan cost

If you already have a Premium or App Service plan, that means you are already paying for the plan, so you can add multiple function apps to the same plan to optimize the cost to some extent. This will eliminate the cold start as you have the compute resources pre-provisioned. At the end of the day, all this boils down to your requirement. If you have simple functions that are not resource-intensive or time-consuming then serverless is the right choice. On the other hand, if you need dedicated compute resources for long-running resource-intensive functions, go for pre-provisioned instances. Moreover, you can further reduce the cost by buying a savings plan for 1 year or 3 years, which gives a discount of up to 17% on the total cost.

- **PoC deployments**: If you are starting to build your solutions in Azure, it is better to start with PoC deployments rather than directly treating the setup as production. Such deployments will help you understand drawbacks and issues with your architecture design, find any single points of failure, and perform some load testing to verify the performance of the application. You can find hundreds of architectures to kick-start your design on the Azure Architecture Center (https://learn.microsoft.com/en-us/azure/architecture/). For example, you want to design a 3D rendering video solution but are not sure about the services and the architecture. You can go to the Architecture Center and search for that, and you will see reference architectures, as shown in the following screenshot:

## Browse Azure Architectures

Find architecture diagrams and technology descriptions for reference architectures, real world examples of cloud architectures, and solution ideas for common workloads on Azure.

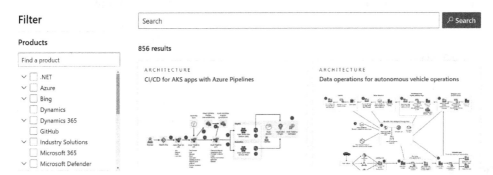

Figure 3.10 – Reviewing reference diagrams

Opening the architecture will show you the complete data path and the services used in the architecture. You can tweak this or use the same for your cost estimation. Apart from the Azure Architecture Center, you can also leverage the Azure Tech Community (`https://techcommunity.microsoft.com/t5/azure/ct-p/Azure`), where you can start a discussion or review architectures shared by community members.

- **Use managed services**: We know that the cloud offers IaaS, PaaS, and SaaS solutions. In IaaS, only the infrastructure will be provided by Azure, and the rest of the management tasks such as updating the operating system, patching, and configuration are the customer's responsibility. IaaS offers more control over the infrastructure but comes with more administration overhead. Microsoft's recommendation is to use PaaS or SaaS services whenever possible; the cost of running the service and maintenance is included in the service cost. Most organizations follow the *lift-and-shift* approach, where they migrate their on-premises VMs as VMs to the cloud. If you have a legacy solution that requires custom binaries and dependencies, then it makes sense to stick to IaaS. However, if you want the full cloud experience with less management overhead, PaaS or SaaS would be the right solution. In this way, we can modernize our infrastructure and get the best from the cloud. To give an example, if you have an ASP.NET application running on a VM and would like to integrate with Azure **Active Directory** (**AD**) authentication, you need to use the **software development kit** (**SDK**) and write code to make the authentication work. With App Services, a PaaS solution is just a matter of a few clicks to integrate with Azure AD authentication, and the best part is there is no code change required.

With that, we have completed the design checklist for cost optimization. Once we have the design done, the next item in the pipeline is provisioning whatever we designed. Now, when it comes to provisioning, there are hundreds of services, and it's not easy to describe the provisioning checklist for each service. Here we will switch to Microsoft documentation which gives us a checklist for each service type. You can refer to the provisioning checklist at `https://learn.microsoft.com/en-us/azure/architecture/framework/cost/provision-checklist`. You can refer to the checklist based on the services that you are planning to provision.

After provisioning the resources, we need to monitor them before we optimize them. Monitoring the resources will give a clear understanding of the utilization and help us set up the baseline cost. On that note, let's discuss the cost monitoring checklist.

## Cost monitoring checklist

Once the resources are provisioned as per the provisioning checklist, we cannot straight away optimize the cost. We need to build data before we can optimize the resources, and this data is collected by monitoring the workloads. Several strategies can be used to monitor cost in Azure, let's discuss these in the following sections, starting with data collection.

### Data collection

Azure provides different data sources for you to collect cost-related information. In Azure, it is very transparent, and you can understand exactly where each penny is going. We can start with **Azure Advisor** and Azure Advisor scores. Azure Advisor is your personalized cloud consultant, which gives customized recommendations based on your environment. The recommendations are given on **Cost**, **Operational Excellence**, **Performance**, **Reliability**, and **Security**. Ever heard of these categories? Yes, these are recommendations based on the pillars of the WAF, which have been discussed in previous chapters. We will be talking about Azure Advisor recommendations when we cover the respective pillars, for now, we will focus on the **Cost** recommendations. In Azure Advisor, we have **Advisor Score**, which is generated by Azure based on the number of recommendations you have. These recommendations can be postponed or dismissed if needed. The following screenshot shows recommendations generated by Advisor and the score:

Figure 3.11 – Azure Advisor recommendations

Once you have deployed the resources in Azure, the telemetry of these resources will be collected by Azure Advisor to generate recommendations for resizing and buying savings plans. Furthermore, Azure Advisor will give recommendations to purchase reservations by reserving the resources for 1 or 3 years with potential savings in cost up to 70%.

Another tool that we can use for data collection is Azure Cost Management. It provides a single glass pane to analyze your cost, set up budgets, and forecast the cost. You can scope your visibility to different levels, such as the billing account (if you are on EA or **Microsoft Customer Agreement** (**MCA**)), management group, subscription, and resource group. The result can be filtered with the help of filters, and you can add multiple filters and further group it based on different dimensions such as service name, service tier, meter category, and so on. The following screenshot shows the usage of filters in Azure Cost Management:

Figure 3.12 – Azure Cost Management

The following are the legends:

- **Yellow**: This refers to **Scope**; here the selection is the **Subscription** scope. In other words, you will see all costs under the subscription. If you scope the view to **Management Group**, then you will see all costs incurred under the management group.

- **Red**: By default, Azure Cost Management loads with the **AccumulatedCosts** view. Let us say you add multiple filters to find the cost related to your business unit. Once you refresh the webpage, all the filters will be reset. If you want to periodically check the cost with the same filters, you can save it as a view, and you can switch views while preserving your preferences. Furthermore, you can see filters. Here we have filtered for the cost of VMs in India Central.

- **Green**: Data can be further grouped with the help of **Group by**; here it has been filtered with the **Meter** dimension, and you can see the cost meters are **B2ms** and **B1ls** (VM sizes). You can control the granularity of the view; currently, we are looking at the cost of the last 12 months. We switched the granularity to **Monthly** so that we can see the monthly cost of VMs grouped by **Meter**. Other granularities include **Accumulated** and **Daily**. For a longer period, such as 12 months, you will not get daily data, but if you are looking at data from the last 2 months or so, then you will have the daily granularity as well. Finally, towards the far right, you can see **Column (stacked)**; this controls how your graph should look. Apart from the current choice, you can convert the graph to an area chart, line chart, column (grouped), and even a table. The default selection when you load Cost Management will be **Area**.

- **Violet**: In addition to the graph in the upper half of the screen, we have donut graphs that you can customize. In *Figure 3.12*, you can see each of the donut graphs has a dimension name (**Service Name**, **Location**, **Resource Group Name**) given at the top with a drop-down arrow. You can change this to any dimension such as **Meter**, **Meter category**, **Meter subcategory**, and so on to display the breakdown in the donut graph. When you save the view in Azure Cost Management, all preferences, including the selection in the donut graph, will be saved for future access.

If you are new to analyzing costs in Azure Cost Management, you can get a better understanding by reviewing this link: `https://learn.microsoft.com/en-us/azure/cost-management-billing/costs/quick-acm-cost-analysis`.

Azure Cost Management can be used to forecast the cost for the next 12 months. In the following screenshot, we can see the forecasted cost for the rest of February 2023:

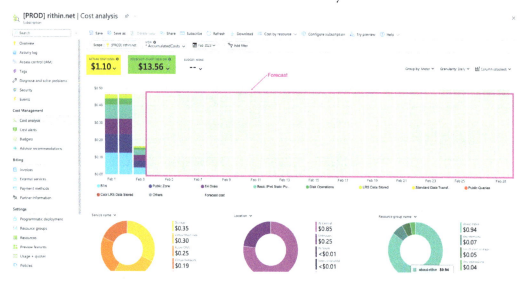

Figure 3.13 – Forecasting cost

The preceding screenshot clearly shows the **ACTUAL COST** (pre-tax) and the **FORECAST** cost. If we don't change any configuration by the end of February 2023, the expected charge is $13.56. All costs you see in Azure Cost Management are the pre-tax amount; based on your local tax authority, tax will be added once the invoice is generated.

If you have a multi-cloud strategy in your organization, you can connect your AWS cost to Azure Cost Management and analyze the cost we perform for Azure. In certain organizations, they prefer to build their own custom tools to analyze the cost; Azure Cost Management offers a REST API, which you can use to integrate with your applications. If you already have a **Power BI** license, then using the Power BI connector, you can ingest the cost of your Azure environment to Power BI and create rich visuals. In this section, we are not touching on other features such as export and budget, but we do cover them in this checklist, as discussed in the *Budgets and alerts* section.

Now that we know how to collect the cost telemetry of your resources, let's see how we can build a good reporting mechanism with the help of Azure resource tags.

### Resource tags

With the help of resource tags, we can add labels or metadata to our resources, which will be sent to the Azure billing system. Since this data is available in the billing system, we can populate these tags in Azure Cost Management, which can be used to filter data based on these tags. Tags are key-value pairs and you can have up to fifty key-value pairs per resource. There are certain key points that must be noted:

- Not all resources support resource tags. Even if they support it, not all of them send the tagging data to the billing system. If the data is not there in the billing, that means we cannot use that in Azure Cost Management. You can review the tag support documentation at `https://learn.microsoft.com/en-us/azure/azure-resource-manager/management/tag-support` and determine whether the resource tag is supported in the cost report. If you are not able to see the tag for a tagged resource in Cost Management, it is most likely that the resource does not send the tag data to the billing system. Before you troubleshoot, it is better to review the documentation.

- In Azure, we see inheritance from a higher level to a lower level in many scenarios. For example, if we add a **role-based access control** (**RBAC**) assignment, a policy assignment, or a resource lock at the resource group level, that gets inherited by all resources that are part of the resource group. However, in the case of a tag, by default, it's not inherited from higher levels to lower levels. Nevertheless, you can use policies to inherit the tags at the subscription scope or resource group scope to all resources that come under it. At the time of writing this book, there is a preview feature that you can enable to automatically inherit tags without the need for a policy; let us wait for that one to come in general availability to test in our production environment.

- Azure doesn't charge you for resource groups; if you tag at the resource group level, that will not be sent to the billing system unless you are inheriting the tags to the resources using policies. Tags at the resource group level can only be used for logical grouping and classification.

- The maximum number of characters supported for a tag name is 512, and the tag value length can be up to 256 characters. Storage accounts are an exceptional case, where the tag value and tag name can only support up to 128 and 256 characters, respectively. Symbols such as <, >, %, &, \, ?, \ cannot be added to the tag value.

- Tags are only supported for resources deployed using Azure Resource Manager, classic resources deployed using Azure Service Manager don't support tags.

While we are provisioning resources, it is not mandatory to fill in the tagging information; you can skip it and deploy the resource. If we do so, the resource becomes untagged, and we cannot track it using tags in Cost Management. However, Azure Policy can be used to enforce the tags so that users need to add the tags when they deploy the resource. Built-in policies are available for you in Azure based on your tagging requirement, and they can be found at `https://learn.microsoft.com/en-us/azure/azure-resource-manager/management/tag-policies`. To give an example, with the help of the **Require a tag on resources** policy, we can enforce a tag called **BillingDepartment** to the resources in the **DNS-domains** scope (you can choose any scope based on your environment) resource group, as shown in the following screenshot:

Figure 3.14 – Policy assignment

After the assignment, the policy takes around 10–15 minutes to start working in your environment. Trying to deploy a virtual network (or any resource) without the **BillingDepartment** tag will fail during the validation of the resource deployment, as shown in the following screenshot:

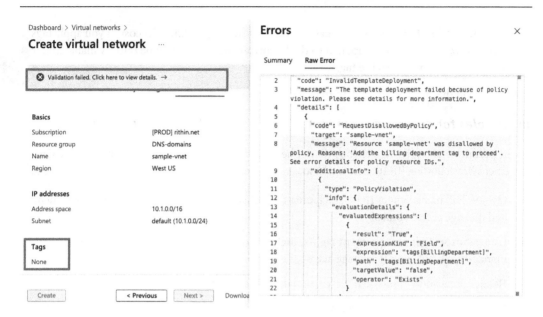

Figure 3.15 – Enforcing tags using policy

If we look at the raw data of the validation error, we can clearly see that the validation has failed because of the missing tags.

You can add multiple policies and create an initiative to assign the policy in one go for better management and compliance checking. Let us say you managed to tag your resources using the tags, then you can use this in Azure Cost Management and filter the cost, as shown in the following screenshot:

Figure 3.16 – Filtering data using tags

In the preceding screenshot, you can see that we are using two tags to filter the cost. If you apply tags to a resource now, that will not be applied to the historical data, and it will be only available under the tag from the time you added the tag.

The next item on the checklist is built-in roles for cost.

### Built-in roles for cost management

Without permissions on the scope, you will not be able to view the cost information in Azure Cost Management. The following RBAC roles can be assigned to your identities for accessing cost management:

- **Owner**: Full management rights, including the rights to view costs, manage cost management, and manage access control

- **Contributor**: Full management rights, including the right to view costs and manage cost management, but cannot manage access control

- **Reader**: Read-only rights, can view cost data but cannot make any changes

- **Cost Management Contributor**: Can view costs and manage cost management

- **Cost Management Reader**: Can view costs but cannot make any changes

While assigning roles to identities, it is better to follow the *principle of least privilege*. For example, if an employee from the finance team would like to view the cost for their Azure environment, there is no need to assign an *Owner* or *Contributor* role; instead, go for the *Cost Management Reader* role. Having an *Owner* role will make the user privileged and lead to issues if the account is compromised. As a best practice, all members of the team should have the *Cost Management Contributor* role added to their respective scope, which enables them to manage costs, build budgets, and export the cost data. On the other hand, *Contributor* business stakeholders can have the *Cost Management Reader* role to track the current costs, as they are not involved in cost analysis or budget creation, which will often be done by the team members.

Nonetheless, we can always create custom roles with tailormade permissions suiting your environment. The key takeaways are to always follow the principle of least privilege, and all stakeholders should have visibility of costs. If your organization signed up for Azure EA, then to view the cost of the entire organization, there are separate roles such as Enterprise Administrators and Department Administrators. These roles are managed from a different portal called the **Azure EA Portal**. With the recent changes in the Azure portal, you can manage the roles from here as well. You can review the EA role documentation at `https://learn.microsoft.com/en-us/azure/cost-management-billing/manage/understand-ea-roles` for more understanding.

With that, we will shift our focus to the next topic: budgets and alerts.

## Budgets and alerts

As mentioned earlier in this chapter, budgets bring accountability for the team to ensure that their cloud spending stays within the approved limit. Azure Cost Management offers the option to set up multiple budgets targeting subscriptions, resource types, tags, and lots more. We can navigate to Azure Cost Management and create a budget, as shown in the following screenshot:

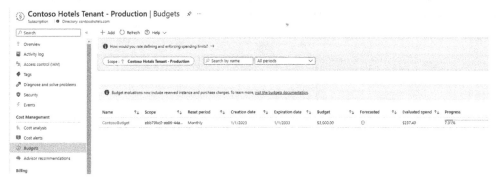

Figure 3.17 – Budgets in Azure Cost Management

Budgets can be created by clicking on the + **Add** button. This will take us to the **Create budget** page, where we can add filters as we have done in Cost Management. We can specify the budget name, start date, end date, and the reset period. Further, we need to provide the budget amount; Azure is intelligent enough to suggest the budget amount based on historical usage. In the following screenshot, we are setting up a budget for VMs in East US called US-East-VM-Budget, which will reset **Monthly** with a defined start date and end date, and have a budget of $300 based on the **$260** budget suggested by Azure:

Figure 3.18 – Creating budgets

As you can see in the preceding screenshot, a red line will be shown in the graph representing the budget we set. Clicking on the **Next** > button will take us to the next screen, where we can integrate alerts with action groups. With action groups, we can set up different notifications and action preferences such as email, SMS, Logic Apps, automation runbooks, webhooks, ITSM, and Azure Functions. There are numerous possibilities with action groups starting from simple email notifications to complex automation workflows. We will be able to set up multiple thresholds and integrate with different action groups to trigger different actions when the threshold is reached, as shown in the following screenshot:

Figure 3.19 – Setting thresholds and alerts

By clicking **Create**, your alert will be created, and you will be notified if you are crossing the threshold. Having multiple thresholds will help us track the cost more efficiently and effectively.

With that, we are moving on to cost analysis and cost reviews.

### Cost analysis and cost reviews

When we are analyzing cost, make use of filters and analyze from all scopes to get a better view. If we look at the cost from the subscription scope or management group scope, the insights that we can gather will be less. The efficient use of filters will help you to drill down the cost and learn the utilization better. We need to use the dimensions such as meter category, meter subcategory, meter, location, and so on to refine the data. The following screenshot shows an example of filtering cost data:

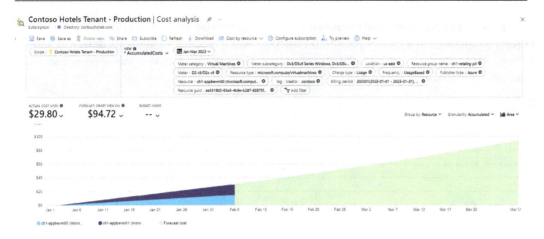

Figure 3.20 – Filtering data in Cost Management

As mentioned earlier, the cost review is particularly important to understand the cost trend. Reviewing costs will help us to understand the deviation from the previous term and to forecast our future spending. Teams need to conduct cost reviews periodically to make sure that they achieve the organizational requirements and that the expense is within the budget allocated for them.

The next item on the checklist is anomaly detection.

## Anomaly detection

Cloud costs are predictable, but sometimes small misconfigurations may lead to surprises in the bill. If we take VMs, for example, the cost for the compute is fixed, which we call the unit price calculated per hour. This makes it easy for us to calculate an accurate cost as we can multiply the unit price by the number of hours consumed. Nevertheless, there are certain metrics that we cannot accurately calculate, for instance, bandwidth. You will be charged for the bandwidth based on the egress or ingress traffic per GB. We cannot say that a service will only consume 5 GB or 10 GB, as it may vary based on the amount of traffic. We need to monitor the historical usage of bandwidth and understand the anomaly, if there is one. We need to monitor for anomalies in the following scenarios:

- Bandwidth cost
- Storage transactions
- Function executions
- CPU utilization and performance bottlenecks

The idea is to recognize these variations that contribute to the cost and include that in your budget if there is a business justification. Azure Cost Management has added an **anomaly detection tool**; however, this feature is in preview at the time of writing this book. You need to enable preview features, as shown in the following screenshot:

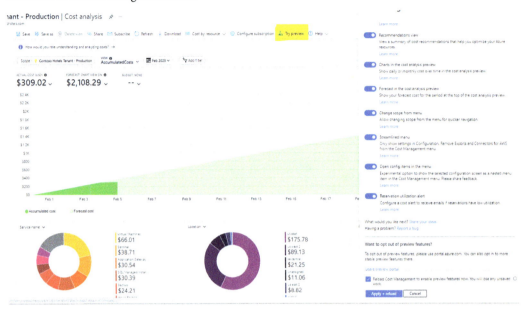

Figure 3.21 – Enabling preview features

Once we apply and reload the configuration, then a new service called **Cost Analysis (preview)** will be registered and you can search for that from the Azure portal. There are multiple views available in the preview portal, and we need to select the **Resources** view to see the insights:

Figure 3.22 – Cost insights

With the help of insights, we can see the daily run rate and any anomaly in the usual run rate. Though this feature is in preview, you can still use it, and once the preview features are enabled, it takes at least 24 hours to build the insights. Once this feature is available in general availability, this will be a game changer helping customers to comprehend anomalies in their environment.

Moving on to the last item in the checklist: visualization.

### Visualization tools

Most business stakeholders are not familiar with the Azure portal and how to analyze costs in Azure Cost Management. It is a great tool; however, proper training should be given to get acquainted with it. To find a common ground for everyone to analyze costs without the need to go to Azure Cost Management, we need visualization tools. The following are the different methods we can use to create visualizations for your cost data:

- **Azure Cost Management Power BI connector**: In the Power BI desktop, we can use the Azure Cost Management connector to ingest data from Azure Cost Management into Power BI. This connector can be used by direct MCA and EA customers only; other offer types are not supported at this moment. The connector uses Cost Management APIs to ingest the data, and multiple tables are available for ingestion. You can refer to `https://learn.microsoft.com/en-us/power-bi/connect-data/desktop-connect-azure-cost-management` to set up the connector. You can develop the visuals locally and publish them; however, publishing requires a Power BI license.

- **Azure Cost Management Power BI application**: The Cost Management connector requires you to build your own visuals once the data is ingested. If you are not familiar with Power BI, then it will be a challenge. A set of predefined visuals and charts, which are ready to use, are shipped with the Cost Management app. Like the connector, the app only supports direct MCA and EA customers only. In order to use the app, you need a Power BI license. You can refer to `https://learn.microsoft.com/en-us/azure/cost-management-billing/costs/analyze-cost-data-azure-cost-management-power-bi-template-app` to configure the app.

- **Third-party solutions**: If you don't have Power BI and are already using other visualization tools, you can integrate with Cost Management APIs to pull the data. You can refer to `https://learn.microsoft.com/en-us/rest/api/cost-management/` for more details on the REST APIs.

With that, we have concluded the monitoring checklist. The data we collected by monitoring the cost will be used for cost optimization. On that note, we will start the optimization checklist.

## Cost optimization checklist

This is the last checklist we have for this chapter. We started with the design checklist, where we covered the cost model checklist, architecture checklist, and monitoring checklist. The lessons we learned will be used for the optimization checklist. The following key points will be considered for cost optimization:

- **Underutilized resources**: We need to review the CPU and memory utilization of the workloads to understand whether they are underutilized. If the number of workloads is large, it's not an easy task to navigate to each machine and review the metrics. This is where Azure Advisor is going to help you. Advisor will constantly evaluate your resources, and based on the historical usage, Advisor will suggest whether you need to resize the VM to a smaller size or completely shut down the VMs. This will make your life easier, and you can easily optimize underutilized VMs (`https://learn.microsoft.com/en-us/azure/advisor/advisor-reference-cost-recommendations`).

- **Cost reviews**: Cost optimization is a continuous process rather than a point-in-time activity. We have already seen how we can conduct cost reviews using different monitoring solutions, jot down your findings from the periodic cost reviews, and act on those findings.

- **Azure Reservations and savings plan**: Azure Reservations and Azure savings plans are cost-saving techniques where customers can make a commitment of one year or three years for their resources to get a discounted price. Reservation purchase requires you to provide the **stock-keeping unit** (**SKU**) and region, and if you need to change this later, you have to either cancel or exchange the reservation. A savings plan, on the other hand, doesn't require you to specify an SKU or region, and it will be applied to the maximum discounted resources, and you can switch to any resources regardless of the SKU and region. You can refer to the Microsoft documentation to learn more about Azure Reservations (`https://learn.microsoft.com/en-us/azure/cost-management-billing/reservations/save-compute-costs-reservations`) and savings plans (`https://learn.microsoft.com/en-us/azure/cost-management-billing/savings-plan`). You can view recommendations for purchasing Azure Reservations from Azure Advisor.

- **Azure Hybrid Benefit**: If you already have Windows, Linux, or SQL licenses purchased from your volume licensing agreement or licensing partner, then you can bring those licenses to Azure use with supported PaaS services and VMs. In this way, you don't have to pay for licensing in Azure, and that will help you save costs (`https://azure.microsoft.com/pricing/hybrid-benefit`).

- **Azure Dev/Test subscriptions**: We have already covered this point earlier in this chapter; nevertheless, make use of Dev/Test subscriptions for your non-production subscriptions as the Dev/Test cost is always less than the production rate (`https://azure.microsoft.com/offers/ms-azr-0148p/`).

- **Autoscaling**: Define scale-in and scale-out policies to handle workloads that have varying demands. Based on demand, the scale-out rule will be able to increase the number of instances, and once the demand is low, the scale-in rule will decrease the number of instances back to the minimum number of instances you have configured. In this way, you don't have to pre-provision a fixed number of instances and pay for them, the rule will take care of the autoscaling, and you only pay for the extra instances during the scale-out period.

- **Revisit your architecture**: Over time, your architecture design might require changes to accommodate business requirements. These changes include altering tiers of services, changing the hosting model, changing the data store, and so on. The following are key areas that you can revisit:

  - **Choosing the storage access tier**: Storage offers hot, cold, and archive tiers, which play a major role in the retrieval cost and data storage costs. If your data needs to be accessed very frequently, then it makes sense to move the storage to the hot tier as the data retrieval cost is very low in this tier, while the data storage cost is high. If the data is not accessed frequently, keeping the data in the hot tier will increase the storage cost, then it's better to move from the cold tier to the archive tier based on the frequency of access.

  - **Choose the right data store**: There are different relational and non-relational data stores available in Azure. Each of these stores comes with its own pricing tier. For example, if you started with the vCore purchasing model of Azure SQL database and later realized that it is expensive, then consider moving to the **database transaction unit (DTU)** purchasing model.

  - **Spot VMs for low-priority workloads**: With the unused capacity in the Azure data center, you can deploy **Spot VMs**, recommended for low-priority and development workloads. Though they are cheaper than the regular VMs, whenever Microsoft needs this unused capacity back, the VMs will be evicted. Due to this reason, we cannot use this for production workloads; however, using spot VMs for low-priority workloads can reduce the compute cost.

  - **Redundancy and data transfer**: Understand the charges for bandwidth in Azure, which is based on the billing zones. If your workload does not require a geo-redundant service, consider switching to a zone-redundant or locally redundant service.

  - **Use managed services wherever possible**: Evaluate the cost of maintaining infrastructure versus the cost of using PaaS or SaaS services. If the cost is less for managed services, consider replacing IaaS with PaaS or SaaS.

By following these considerations, you can optimize the cost of running workloads in Azure. Once again, optimization is not a one-time task or something that you perform when you see a spike in usage; you need to consider optimization as an iterative process driven by the cost reviews and monitoring of your workloads. Too much of anything is the beginning of chaos; if you over-optimize, that will lead to certain tradeoffs. Before concluding this chapter, let us quickly cover the tradeoffs that you have to be concerned about while performing cost optimization.

# Tradeoff for cost

The WAF is a framework that covers a set of best practices touching the pillars of cost optimization, namely, operational excellence, performance efficiency, reliability, and security. One thing you have to remember is an ideal solution does not equate to a low-cost solution; there will be tradeoffs with other pillars such as operational excellence, performance efficiency, reliability, and security. The question that you need to ask yourself is: What is the top priority for your business? Is it low cost, no downtime, or high performance? Based on your answer, there will be compromises that you have to accept. Let us cover the tradeoffs between cost and each pillar of the WAF:

- **Cost versus reliability**: Reliability and cost are inversely proportional. If you want to implement high availability and disaster recovery for your workloads, that will lead to additional costs. The rule here is to verify whether metrics such as SLAs, RTOs, and **recovery point objectives** (**RPOs**) are met as per business requirements if you optimize the cost. For example, your solution requires Azure Storage with 99.999% SLA and is currently using geo-redundant storage. You noticed that cost for storage is high and plans to move to zone-redundant storage. Obviously, this will reduce the cost, and at the same time, the SLA for Storage will be dropped to 99.99%. This will be an SLA breach as the solution requires 99.999% and with your optimization, it decreased to 99.99%; this will not meet business requirements. In these kinds of scenarios, we need to focus on other parameters such as the access tier, and see whether there are opportunities for optimizations.

- **Cost versus operational excellence**: In operational excellence, we focus on monitoring and automation, which might increase the cost in the initial days but, over time, reduce the cost. Remember that we discussed in the *Cost monitoring checklist* section that setting up automation and budgets can, in turn, help with efficient cost management.

- **Cost versus performance efficiency**: The more performance you need, the bigger the machines that you need to deploy. We can manage this to an extent by evaluating the need for pre-provisioned versus serverless; moving to serverless will help reduce the cost, but you have to deal with the cold start. Similarly, deploying to a cheaper region can reduce cost, but then you need to worry about the latency. You need to find a balance between performance and cost based on business requirements.

- **Cost versus security**: If you need additional security, then you need to deploy **network virtual appliance** (**NVA**)/Azure Firewall, enable Defender for Cloud – Standard plan, enable **Distributed Denial-of-Service** (**DDoS**) Protection – Standard plan, enable Sentinel, set up Web Application Firewall, and so on. These additional security solutions will have an impact on the cost. Try using native security solutions such as a **network security group** (**NSG**), service endpoints, private endpoints, and so on at the beginning. If the business demands extra security, then we can add premium features.

As we move on to other pillars of the WAF, we will cover the tradeoff on those respective pillars. The bottom line is, everything boils down to business demands. If you are ready to sacrifice performance, you can make great savings, but the question would be how that will impact the business. Similarly, we can think about other pillars. The WAF is not only about cost optimization, we need to think from the perspective of other pillars as well. You will learn about these in the upcoming chapters.

## Summary

In this chapter, we touched on the first pillar of the WAF: cost optimization. We started with an introduction to cost management and the need for it. As we progressed, we explored the design principles that we need to consider before moving our workloads to the cloud. After that, we covered the checklists. The checklists are crucial for designing, provisioning, monitoring, and optimizing cost in Microsoft Azure. The first step is to understand the design checklist and embrace the cost model and architecture design aspects. Keeping the design checklist in mind, we need to provision the resources required for our workload. We cannot start the optimization at this point because we do not have enough data or telemetry to make a decision. This is where the monitoring checklist comes into the picture, where we learn to collect the data and visualize it in a meaningful way. Once we have telemetry, we can start optimizing the workloads. While optimizing the workloads, remember that there are tradeoffs that we need to consider.

As we move on to other pillars of the WAF, you will learn how closely these are interconnected. On that note, let us move on to the next chapter: *Achieving Operational Excellence*.

# 4

# Achieving Operational Excellence

**Operational excellence** is the second pillar in the **Well-Architected Framework (WAF)**. In the previous chapter, we covered the first pillar, cost optimization, and we also covered the tradeoffs between the pillars. There is no specific order that states that operational excellence is the second pillar of the WAF. The pillars are arranged in order to provide the best reading experience. The key principle that operational excellence brings to the table is that all deployments must be predictable and dependable. Though it is a bit ambiguous, by the end of this chapter, you will be able to understand how we can bring predictability to our deployments.

We will adopt a similar approach that you have seen in the previous chapter; we will start with the introduction to the pillar, then the design principles. Further, we will discuss automation and release engineering. Let us start with an introduction to operational excellence.

## Introducing operational excellence

Operational excellence focuses on a set of operations and processes that we need to implement for our applications running in production. As mentioned in the opening paragraph of this chapter, the idea is to make deployments **reliable** and **predictable**. Reliable as in eliminating the chances of human error; as you know, when we perform tasks manually probability of human error is high. In order to reduce this human error, we need to leverage **automation**. With the help of automation, we are able to perform a series of steps without the need for human intervention. Since we rely on automation for the deployment, the outcome is predictable. When it comes to deployment, we need to make sure our **release management** is optimized for pushing new features and bug fixes seamlessly. At the same time, we should be able to roll back to the last known good configuration if the deployment did not go as expected.

With the operational excellence assessment, we will be reviewing if all these points are covered. The following topics are part of the operational excellence pillar:

- **Application design**: Keeping the DevOps principle in mind, guidance will be provided on how to design, build, and implement workloads in Azure.

- **Monitoring**: Once the workloads are implemented, we need to monitor the performance metrics and logs of these workloads. If these workloads were in an on-premises data center, we could extract the information from the hypervisor if something goes wrong. However, in the case of Azure, it will be a remote data center, and customers will not have access to the underlying infrastructure. Customers should focus on implementing monitoring for the resources deployed on Azure rather than worrying about the underlying infrastructure. Microsoft will take care of the underlying infrastructure and its monitoring.

- **Application performance monitoring**: With the help of tools, we need to monitor the performance and availability of our application. In Azure, we can use Application Insights to fulfill this requirement.

- **Code deployment**: We need to ensure that the code is deployed quickly and seamlessly. With the help of DevOps, we can ensure that bug fixes and patches are deployed using release management.

- **Infrastructure provisioning**: As we discussed earlier, repetitive tasks need to be automated to avoid human error; this is applicable to infrastructure provisioning as well. With the deployment automation or **infrastructure-as-code** (**IaC**), we will automate the infrastructure provisioning.

- **Testing**: Testing needs to be performed to ensure that the application is free from any errors or mistakes before it is pushed to production. If we miss testing, the application will be shipped with errors or mistakes, impacting our users.

With the help of Azure policies, we can enforce resource-level restrictions to ensure that the operational excellence pillar requirements are enforced on our workloads. To give an example, we can ensure that only approved **virtual machine** (**VM**) **stock-keeping units** (**SKUs**) are used in production for optimal excellence with the help of the **Allowed Virtual Machine SKUs** policy. Similarly, based on your organizational requirements, you can generate additional policies. In *Chapter 3*, we saw the cost recommendations in Azure Advisor, and you were able to review the recommendations related to operational excellence. The following screenshot shows the **Operational excellence** recommendations generated by Azure Advisor:

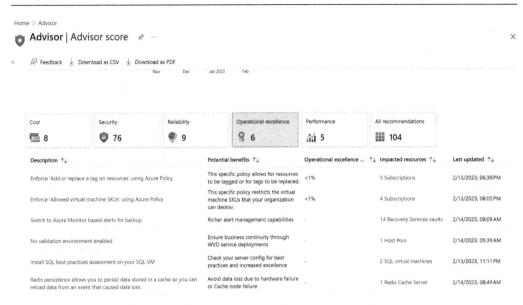

Figure 4.1 – The operational excellence recommendations

If you are not able to comprehend the recommendations, clicking on a recommendation will take you to a page where you can see its description, the impacted resources, recommended actions, and other subscription-related information:

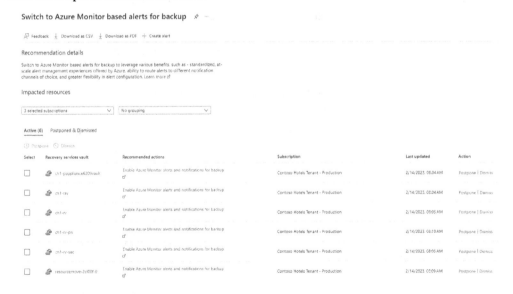

Figure 4.2 – Reviewing recommendations

Furthermore, if you would like to use the **Postpone** or **Dismiss** recommendations, you can do that with Azure Advisor. If you use **Postpone**, the recommendation will be removed from the list and resurface after the specific time you configure. During this time period, you can decide whether you want to complete the recommended actions. On the other hand, **Dismiss** will remove the recommendation from the list, and it will not resurface again. Nonetheless, you can always switch to the **Postponed & Dismissed** tab and view the removed ones and reactivate them if required.

On that note, we will move on to the design principles.

# Design principles

As we have seen in the case of cost optimization, we do have design principles for operational excellence as well. By adopting these principles, we can ensure that our workloads are aligned with the operational excellence outcomes. In order to attain high proficiency in operations, we should consider improving the following factors:

- **Software development**: There are multiple ways of developing software, and the method we chose plays a critical role in the success of the project. Examples of these methodologies are Agile, Waterfall, DevOps, Rapid Application, and so on. Each of these methodologies has its own pros and cons for varied reasons, and we will not be covering these, but the idea is that adopting one of these adopted methodologies can have a positive impact.

- **Software deployment**: Software deployment can be performed via a manual push, or we can go for **continuous integration/continuous deployment** (**CI/CD**). Manual deployments are prone to human error, and we want to eliminate this, considering CI/CD would benefit us by continuously deploying new code to production whenever the developers commit the code to the source control.

- **Software operation**: We need to analyze the usage patterns to understand the software operations. This is where we decide the high availability strategy of the application.

- **Software maintenance**: Maintenance includes updating code, patching, and bug fixing. These tasks should be performed seamlessly with the ability to roll back in case of failure.

These factors are all related to software, and it is equally important that we bring in a team culture that comprises the following:

- **Trialing and expansion**: It is important that we encourage the team to adapt to the growth mindset rather than taking a fixed mindset. Technology is growing every day; before virtualization, we used to buy physical hardware and host our workloads. Though virtualization was introduced in the 1960s, it was not widely adopted until the early 2000s. With virtualization, we learned that we can host multiple machines within the same host server. Now this has evolved into the world of containers and **platform-as-a-service** (**PaaS**) solutions. Instead of sticking to legacy methods, the team should be ready to experiment with innovative technologies and update their environment.

- **Making the current process efficient**: As we move to the cloud, there are diverse ways to improve its efficiency. The team should adopt features such as monitoring, high availability, and automation to enhance the current process that is being used.

- **Incident response plans**: As we discussed in *Chapter 2, Distinguishing between the Cloud Adoption Framework and Well-Architected Framework*, monitoring is an iterative process that may require amendments as the business requirements change. The team should have a proper strategy defined to acknowledge incidents in case of one. These processes include business continuity plans, disaster recovery plans, workload scaling plans, and so on.

The principles that we will cover can be used as a baseline to evaluate our applications deployed on Azure. Let us start with the principles.

## Optimize the build and release process

Improving the software build and release process is one of the major optimizations we need to accomplish in the operational excellence pillar. Though it may sound like a repetition of what we discussed at the beginning of the chapter, we need to go through the relevant points again to cover the design principles. Let us cover them one by one:

- **Deploy using IaC**: With the scale of infrastructure, managing infrastructure manually can lead to errors and inconsistencies. Using IaC, we can organize the infrastructure. Modern IaC tools such as **Terraform** have state management, which helps build consistency in the environment. In Azure, we have Bicep as an IaC solution. Bicep does not provide state management as such; however, you can directly view the state of resources in Azure. This helps teams to build infrastructure without worrying about statement management.

- **Ensure CI/CD**: With the help of CI/CD pipelines, we can make the release process flawless and seamless. For example, if you use Azure App Service to host your application, you can integrate with source controls such as Git, GitHub, BitBucket, and so on. Whenever there is a new commit to your repository branch, we can automatically push the code to App Service. This will help to release new code immediately to production. Furthermore, we can make use of deployment slots in App Service to set up CI/CD in your development branch. In this way, we can release production and development code effortlessly.

- **Avoid manual testing**: With the help of automated testing, users can conduct tests even if they are not knowledgeable about the test processes. Using automated testing tools will also give you the advantage of running tests on demand or on schedule. Furthermore, if a part of the test fails due to infrastructure issues, then we can rerun the failed ones instead of running the whole test again. Let us say that if you are using Azure DevOps, you can utilize a test plan for performing automated testing.

- **Adopt configuration-as-code** (CaC): Earlier, we discussed IaC, which can be used for infrastructure provisioning. Similarly, CaC is used to manage workload-specific configuration. It can be quite useful if you are storing your configuration in a source control and configuration migration between environments. IaC and CaC are not mutually exclusive. Though they can work together hand-in-hand, they have their differences. Combining both IaC and CaC in your environment can provide unequaled control over your resources and the ability to avoid configuration drift.

Adopting these all the way through your software development cycle enables you to build the following:

- **Consistency**: With the help of IaC and CaC tools, you can accomplish environment consistency. Adoption of these tools can eliminate errors that can happen during manual processes and configuration drift.

- **Repetition**: Since the infrastructure and configuration are stored as artifacts in source control, we can reuse the code to deploy the workloads repeatedly with the same configuration. As the code remains the same for all deployments, the entire process is repeatable.

- **Early detection of issues**: Adopting automatic testing helps to conduct tests on demand or based on schedule without manual intervention. The test results can be reviewed, and issues can be detected at the earliest opportunity. With the help of CI/CD, we can deploy the code directly to the test environment and verify how the application will behave in production.

Now that you are familiar with the improvements you can make in the build and release process, let us discuss the next principle.

## Comprehending operational health

Monitoring is required to identify issues in operation health and address them before they turn into bigger problems. If you have a robust monitoring strategy, you can see all the telemetry of your workloads, which allows you to troubleshoot and take proactive steps to mitigate issues. In Azure, we have Azure Monitor that can be used to collect telemetry from distinct levels starting from the platform level all the way up to the application level from discrete sources. Key monitoring strategies you need to implement include the following:

- **Monitoring build and release process**: With the help of monitoring the build and release process, we are able to address any issues that can happen in the pipeline. If you are using Azure DevOps, we can integrate with Application Insights and monitoring from other Azure resources. Application Insights alerts can be used to *gate* or *roll back* the deployment until the alert is resolved. Think of monitoring the build and release process as a checklist with a set of items you want to check before you push deployment to production. If you do not pass everything on the checklist, then that is a red flag, and the deployment will be halted immediately. If all checks pass, then the deployment will be pushed to production automatically without any manual intervention. The process is quite simple, which acts like a safety switch ensuring that you meet the prerequisites before deploying to production.

- **Infrastructure health**: Once the deployment is completed, your application will run on the infrastructure. In Azure, we can use different compute solutions to host your application. For example, if you are using Azure VM, you need to monitor the health of the VM to ensure that it is available to serve requests from end users. We need to monitor metrics such as CPU, memory, and so on, and ingest the logs to a Log Analytics workspace. Also, without monitoring in place, we are not able to monitor the state of the machine. For instance, if there is a high CPU utilization, that could impact user experience. We need to add more instances, or if the behavior is unusual, then analyze the logs as well. Furthermore, we can add load balancers and set up health probes to monitor the health of the machine. The bottom line is to ensure that infrastructure is up and running to serve the end users with the help of monitoring.

- **Application health**: Infrastructure monitoring can help you collect metrics and logs from the host; however, what if something goes wrong with the application stack? For example, if your application is returning a 500 error that cannot be monitored from the host level. We need to collect logs from the application stack to understand what went wrong. With the help of **application performance monitoring (APM)** tools, we can monitor the health of our application and resolve the issues. In Azure, we can use Azure Application Insights to collect telemetry from your application. If you incorporate Application Insights into your application, you can get features such as live metrics, availability tests, DevOps integration to Azure DevOps or GitHub, application usage telemetry, failure detection, and anomaly detection.

Microsoft has developed a framework called Azure mission-critical framework concerning health modeling for business-critical and mission-critical applications. You can refer to `https://learn.microsoft.com/en-us/azure/architecture/framework/mission-critical/mission-critical-health-modeling`; this document provides direction and illustrations of health models for your applications. On that note, we will move on to business continuity and disaster recovery.

## Business continuity and disaster recovery

When you design workloads, you need to plan for **business continuity and disaster recovery (BCDR)**. Regional outages are common in the cloud, so it is crucial that we plan to tackle this scenario. The plan needs to be documented so that the team can perform the necessary actions at the time of a disaster. You need to practice failure and recovery exercises for the team to be familiarized with the processes involved. The following methods need to be considered for your BCDR strategy:

- Perform disaster recovery drills on a regular basis to ensure that **recovery time objective (RTO)** and **recovery point objective (RPO)** values are achieved according to the business requirements.

- Conduct chaos engineering tests on the distributed systems to make certain they can tolerate unforeseen outages and disruptions.

- Prepare failures to validate the efficiency of your existing recovery plan and ensure teams are well aware of their responsibilities and the steps that they need to perform in case of a catastrophe.

- As is commonly said, "Correct your mistakes as an opportunity to learn and develop." Whenever you encounter failures, document them; this can be used as a learning for improvising your plan.

- Automate remediation when there is a failure, as this reduces the response time and eradicates the need for manual involvement.

Building a strong BCDR plan will build confidence in your organization because your team knows about the history of failures as they are documented, and at the same time, you have your plans in place to manage the disaster.

Moving on, we will discuss the next principle – accept continuous operations enhancements.

## Accept continuous operations enhancements

Never consider your operations as a fixed tactic; periodically review your plans and bring in process enhancements. Evaluating your operational methods is one of the key elements in aligning operational excellence. The reason for these enhancements is to reduce operational complications and vagueness in your environment.

Consider developing a continuous improvement plan for your organization, which will help you in the following:

- **Process evolution**: Adopt new methodologies and processes in your environment. For example, if you do not have a proper software development process in place, then consider using Agile, Waterfall, or another. This process will help your organization to develop solutions and implement them successfully in a strategic manner. Another example is, say, you are using legacy authentication, where users use their username and password to sign in to applications. This process opens a vector for cyberattack as credentials can be leaked. To avoid this, we need to implement **multi-factor authentication** (**MFA**), which will ensure that users need to pass additional checks before access is granted.

- **Process optimization**: You may have certain processes in place; however, chances are it's not yet optimized. With optimization, we aim to eradicate the inefficiency and gaps in our processes. For example, you need to shut down 20 servers during the weekend to avoid additional costs. From a cost optimization standpoint, this is the right thing to do. Nevertheless, if these servers are stopped manually by user intervention, then it is not an efficient process. We could optimize this with the help of automation, and the servers will be shut down based on the schedule you define without any user interference.

- **Learn from failures**: When we covered the BCDR plan, we saw the importance of learning from failures; this also applies to process enhancements. Let's take the previous example, where we wanted to shut down 20 servers during the weekend. Assume that there is no automation and the administrator would shut down the machines manually. Chances are there the administrator might miss a couple of machines, letting them run during the weekend, resulting in extra cost. This is a failure, and we need to learn from this. The lesson is to not rely on manual processes; it is better to convert them to automated ones.

- **Continuous evaluation**: In order to implement new opportunities and enhancements, you need to continuously evaluate your environment. Processes that may appear efficient at the beginning can be replaced by effective newer methods. To give an example, you are using Kubernetes as a container orchestrator for your containerized workloads. The cluster is running on Azure VMs, comprising one master node and three worker nodes. Since this is an **infrastructure-as-a-service (IaaS)** solution, you need to manage the master and the nodes. During your evaluation, you understand that you can replace this with **Azure Kubernetes Service (AKS)** without the need to manage the master node from your side. If this evaluation was not conducted, then you would have been using the existing process without any change.

Now that you are familiar with the process enhancements, let's discuss the next design principle.

## Use loosely coupled architecture

Currently, we are transitioning from monolithic architecture to loosely coupled architecture. Adoption of these modern architecture patterns, paired with cloud design patterns and advanced deployment strategies, enables teams to build and deploy services to the cloud. The deployment can be performed independently, and if there is a service failure, the impact can be minimized. We do not want to get into the details of these patterns and strategies; nonetheless, if you are not familiar with these concepts, it's recommended that you review these terminologies. To build a loosely coupled architecture, you need to do the following:

- Leverage **modern architectural patterns**, such as the following::

  - Serverless

  - Loosely coupled

  - Microservices

- Combine these with **cloud pattern designs**, such as the following::

  - Circuit breakers

  - Load leveling

  - Throttling

- Integrate **advanced deployment strategies**, such as the following::

  - Canary deployment

  - Blue-green deployment

  - Staggered deployment

Once the teams implement the loosely coupled architecture, they will be able to support, operate, and maintain architectural components independently, provided there are no dependencies involved with other teams. The beauty of loosely coupled architecture is evident in the way the individual components can be maintained without having to stop the entire application.

With that, we have completed the design principles for operational excellence. Consider these principles as your gospels when you develop solutions with operational excellence in mind. Now we will move on to one of the core topics of operational excellence: automation.

## Automating deployments

We have seen the rationale behind using **automation** throughout this chapter. In this section, we will cover the goals, best practices, and types of automation that you can use in your environment. Automation is not a new thing; we have been using automation for ages with the help of scripts. Prepping Windows Servers and Linux servers is a classic example of automation that we used in the past. It has brought in a transformation, has altered the way businesses run, and never ceases to evolve. With automation, we can save a lot of time, and administrators do not need to perform manual tasks. The time saved by administrators can be invested in more productive tasks, which will bring in more business value. A common myth is that automation requires coding, and to an extent, that is correct; however, in Azure, there are out-of-the-box solutions that can perform automated actions and eliminate the need for manual processes without the need for coding, making it available for administrators who do not have coding or scripting background. Using automation, we can accomplish the following:

- **On-demand activation of resources**: In Azure, we have elasticity and scalability; this can be used for the autoscaling of our resources. Previously, we used to increase or decrease the number of instances with the help of manual processes to accommodate the demand. With autoscaling, based on the criteria we set, Azure will increase or decrease the number of instances, automatically making resources available on demand.

- **Agility**: One of the advantages of the cloud is its agility and the ability to create or deploy solutions rapidly. Let us say you want to deploy 10 identical VMs in Azure; if you choose the Azure portal for the deployment, you need to go through the VM creation process 10 times with a higher chance of error. Using automation, we can simply write a PowerShell or Bash script and iterate using a loop 10 times, and the VMs will be ready. Another approach is to use IaC solutions such as **Azure Resource Manager** (**ARM**) templates, Bicep templates, and so on to automate repetitive tasks.

- **Minimize human error**: While configuring repetitive tasks, the chances of human error are very high. Let's take the same example where we need to deploy 10 VMs. With the Azure portal, you have to go through the VM basics, storage, network, advanced configuration, tagging, and so on, 10 times, increasing the probability of error. Automating the deployment is the solution to this problem.

- **Consistency**: Since we are using the same template or script for the deployment, all resources created will exhibit consistency. With the help of these templates, we can achieve repeatable results with very few chances of error.

Let's take a look at the goals of automation.

## Goals of automation

The goal behind automation is to avoid performing repetitive tasks in a manual fashion and eliminate any human error. The beauty of automation is we don't have to reinvent the wheel; all you need is a tool that can convert human action into a script or template, which can be used to perform boring tasks error-free. Ideally, we need to incorporate automation in any task that requires multiple iterations over long periods of time. Apart from error-free actions, increasing the speed of the process is another goal of automation. For example, if you want to create 10 Ubuntu VMs in Azure, it may take 30–45 minutes to manually create those VMs. If we perform the same using Azure PowerShell, all you need is 10 lines of PowerShell code:

```
for ($i = 1; $i -le 10; $i++){
    Write-Host "Creating ubuntu-vm-$i"
    New-AzVm `
    -ResourceGroupName $resourceGroup`
    -Name "ubuntu-vm-$($i)" `
    -Location 'East US' `
    -Image UbuntuLTS `
    -size Standard_B2s `
    -GenerateSshKey `
    -SshKeyName vmsshkey
}
```

Here, with the help of a `for` loop, we are creating 10 Linux VMs, and the entire process is automated and does not require any manual interference. The preceding code block has a simple configuration; if you need to add more customization such as a virtual network, interface, or network security group, then the better approach is to use an ARM or Bicep template. The automated approach can save time, and administrators can work on other tasks instead of deploying these VMs manually.

The preceding sample code uses PowerShell, which can be replaced by IaC solutions. Another automation that the operational excellence pillar describes is the use of CI/CD. Previously, developers used to write the code and evaluate it; once they were ready to ship the code, they reached out to the operations team to push the code to the production servers. The complete process was manual and required effort from both developers and the operations team. We already covered CI/CD earlier in this chapter; with the help of CI/CD, we can deploy code seamlessly as soon as the developers commit the code to the source control. In CI/CD, we also eradicate the need for manual intervention for code deployment and make the code-pushing process hassle-free.

Regardless of automation, there is always a small percentage of toil involved in the process. Toil refers to the operational work linked to a process that is manual, boring, can be automated, and has negligible value. If the percentage of toil is too high, that will impact productivity and slow development. Irrespective of whether it is a planned or unplanned toil, if it constantly disrupts the pace of the project, then it needs to be eradicated. Considerable amounts of toil can impact the job satisfaction and productivity of engineers. If the team finds themselves spending a lot of time on repetitive tasks, they will get bored and dissatisfied. They will not be able to contribute productively to the project as they are overwhelmed with the amount of boring manual work. However, though it sounds counterproductive to automation, a small amount of toil is inevitable. When you design automation, consider avoiding and eliminating toil. If you manage to do so, then engineers can focus on productive tasks and can be more efficient.

To conclude, automation makes any process error-free, time-saving, and productive. Now that you are familiar with the goals of automation, we will learn the best practices linked to automation.

## Best practices for automation

Consider the following best practices while you design automation:

- Always ensure that there is consistency; the more manual processes involved in your environment, the more the probability of human error. Problems such as configuration mistakes, inaccuracies, degraded data quality, and lack of reliability can be caused by these manual processes. With automation, we intend to bring consistency to avoid these issues.

- Having a centralized platform to track issues and bugs. Azure DevOps is an example of this. With this repository, we can collaborate with other teams and aim to create bug-free solutions.

- Complex issues require time to fix and can lead to inefficiency. With the help of automated tests, we can find the issues, verify them, and eliminate them. These tests will ensure that the code is ready to deploy only after passing the checks.

- Raising employee productivity is another best practice. With automation, we can build innovative tools that will save time for engineers. This, in turn, increases job satisfaction, and employees will be more productive as they can utilize their time for other innovations. Once the automation is in place, we need to provide training so that they are familiar with the automated process.

Think of complex boring tasks in your environment and see how you can automate them. After automation is done, you can share that with the entire team and see how happy they are. On that note, we will move on to the types of automation.

## Types of automation

As you embrace cloud transformation, there will be scenarios where you need to perform repetitive tasks, mostly on the infrastructure side. When we deploy solutions repeatedly, the chance of errors is high, and we need to automate the infrastructure provisioning, configuration, and operational tasks.

We already covered IaC and CaC earlier in this chapter. We can rely on these technologies to make our life easier. In Azure, we have multiple technologies available to automate infrastructure deployment, infrastructure configuration, and operational tasks.

We can classify automation broadly into three types:

- Infrastructure deployment
- Infrastructure configuration
- Operational tasks

Let us try to understand each of these in detail.

## Infrastructure deployment

As the name suggests, with this type of automation, we are automating the infrastructure provisioning. We have multiple examples of this type of automation throughout this chapter, where we need to deploy infrastructure repeatedly. The rationale behind this automation is to bring consistency and reliability across our deployments. The primary technology we use for this automation is IaC technologies. Here, we specify the infrastructure components in the form of code, which is used for the provisioning. These technologies use **declarative automation**, where we declare what we want to deploy in the form of code in a sequence, and the provider handles the creation of these resources. As an example, you create an HTML file with all the elements you need and their formatting. Now when we provide this file to the browser, it will display the elements based on the code we wrote. There are different technologies available on the market for infrastructure deployment in Azure; the prominent ones are as follows:

- ARM templates
- Azure Bicep
- Terraform

At the end of the day, all these technologies use their own language to declare the infrastructure, then ARM will read it and deploy the resources we requested. Though this book is not about the pros and cons of these technologies, let us quickly understand the structure of these technologies. We will start with ARM templates.

### ARM templates

The ARM template is the native IaC solution available in Azure, besides Bicep, and is written in JSON format. In our template, we will define the resources that we need to deploy for Azure deployments. Azure has a management layer called ARM, which is responsible for creating, updating, and deleting resources in Azure. When we deploy our template, the resources we request are created with the help of ARM by making the relevant API calls.

ARM templates bring advantages to the table such as parallel resource deployment, modular deployments, validation before deployment, testing, and so on. The following code shows the ARM template for the creation of a storage account:

```
{
    "$schema": "https://schema.management.azure.com/
schemas/2018-05-01/subscriptionDeploymentTemplate.json#",
    "contentVersion": "1.0.0.0",
    "parameters": {},
    "resources": [
        {
            "name": "packtdemostorage",
            "type": "Microsoft.Storage/storageAccounts",
            "apiVersion": "2021-04-01",
            "location": "EastUS"
            "kind": "StorageV2",
            "sku": {
                "name": "Standard_LRS",
                "tier": "Standard"
            }
        }
    ],
}
```

As you can see in the preceding code snippet, we can include parameters and variables to make the deployment more dynamic. For example, if you want to dynamically pass the storage account name every time you deploy the template, then you can create a parameter for that instead of hardcoding the name as `packtdemostorage`. Accommodating these kinds of parameters will make your code reusable without the need to change the template.

If you are interested in writing ARM templates, it is worth reading the documentation available at `https://learn.microsoft.com/en-us/azure/azure-resource-manager/templates/overview` and using the Visual Studio Code ARM tools extension (`https://marketplace.visualstudio.com/items?itemName=msazurermtools.azurerm-vscode-tools`) for a better authoring experience.

Let us move on to the next one on the list: Azure Bicep.

## Azure Bicep

If you look at the ARM template, it is written in JSON, and for many people, JSON is not a friendly language structure. JSON requires it to follow the schema and it is hard for beginners to interpret the code and debug it if they face any issues. Azure Bicep is a replacement for ARM templates if you do not like to write your declarative automation in JSON. Bicep reduces the complexity of the code by using a **domain-specific language (DSL)**. Behind the scenes, when you deploy a Bicep file, the Bicep

**command line interface** (CLI) will convert the file to an ARM template and deploy the resources. Here you do not have to deal with the complex JSON code; the Bicep CLI will take care of that. Treat Bicep as a revision of the ARM template rather than thinking of it as a new language. Obviously, the syntax will be different and simpler than ARM templates, but the underlying functionality and runtime still are the same.

The following code shows a Bicep file that can be used to create a storage account:

```
param location string = resourceGroup().location
resource storageaccount 'Microsoft.Storage/storageAccounts@2021-02-01'
= {
  name: 'packtdemostorage'
  location: location
  kind: 'StorageV2'
  sku: {
    name: 'Standard_LRS'
  }
}
```

You can see how simple the code is compared to the ARM template. Here we can also introduce parameters to remove hard-coded values. If you would like to learn more about Bicep, have a look at the documentation at `https://learn.microsoft.com/en-us/azure/azure-resource-manager/bicep/overview` and use the Visual Studio Code Bicep extension (`https://marketplace.visualstudio.com/items?itemName=ms-azuretools.vscode-bicep`) for a better authoring experience.

ARM templates and Bicep files are native to Azure; however, the next one we are going to cover is a cloud agonistic solution that you can use in any cloud provider such as AWS, **Google Cloud Platform (GCP)**, and so on.

### Terraform

Terraform is another IaC solution that supports the deployment and configuration of infrastructure in many private and public clouds. As mentioned earlier, due to its cloud-agnostic framework, engineers do not need to learn native tools for each cloud; instead, they can use the same syntax for all deployments. Terraform uses a DSL called **HashiCorp Configuration Language** (HCL). Similar to other tools we discussed earlier, Terraform Azure Provider is an abstraction on top of the Azure Management APIs. When we deploy the Terraform file, the provider interacts with the underlying Azure APIs to deploy the resources.

The following snippet shows Terraform code that can be used to create a resource group and storage account:

```
#Create resource group
resource "azurerm_resource_group" "example" {
  name      = "newStorageAccount"
```

```
    location = "eastus"
}

#Create storage account
resource "azurerm_storage_account" "example" {
  name                     = "packtdemostorage"
  resource_group_name      = azurerm_resource_group.example.name
  location                 = azurerm_resource_group.example.location
  account_tier             = "Standard"
  account_replication_type = "LRS"
}
```

If you are interested in learning about Terraform on Azure, then you can refer to `https://learn.microsoft.com/en-us/azure/developer/terraform/overview`.

Now that you understand how automation can be used for infrastructure deployment, let us move on to infrastructure configuration.

## Infrastructure configuration

Once the infrastructure is provisioned, the next thing we need to perform is **configuration management**. For example, you need to create 10 Linux VMs and install Apache, PHP, and MySQL on these servers. You already know that we can create 10 VMs using ARM templates, Bicep, Terraform, or any other IaC solution. We could connect to each of these deployed VMs and install Apache, PHP, and MySQL. However, this will be a repetitive process, and the chances of errors will surface again. Here we need to use infrastructure configuration tools to install these packages programmatically and configure the VMs.

We can classify infrastructure configuration solutions into two types:

- Bootstrap automation
- Configuration management

Let us understand the difference between these solutions, starting with bootstrap automation.

### Bootstrap automation

If you look for the definition of **bootstrap** online, you may come across the following definition provided by the Heathen Knowledge Base (`https://kb.heathen.group/company/design/bootstrap-scene#introduction`): "*A technique of loading a program into a computer by means of a few initial instructions.*" In certain scenarios, we need to run scripts or programs post-resource deployment, and we call this **bootstrapping**. There are several options available for bootstrapping tasks. Let us explore these solutions.

- **Azure VM extensions**: Using VM extensions, you can run post-configuration scripts on your VMs. There are multiple extensions available for VMs that can be used for configuration management, anti-virus, security, and so on. When you write your ARM templates, Bicep, or scripts, you can invoke the extension and run the scripts stored in the Azure storage account or GitHub repository. When the extension is invoked, the extension will download an agent and execute the script within the VM. You can learn more about VM extensions at `https://learn.microsoft.com/en-us/azure/virtual-machines/extensions/overview`.

- **Cloud-init**: For configuring Linux VMs, we can use `cloud-init`. The configuration we add to the `cloud-init` file will be executed on the first boot of the VM. Ideally, the process writes your configuration to a file called `cloud-init.txt`. This configuration will be pushed to the VM, and it will execute the instructions we have added to the file. There are a lot of `cloud-init`-enabled VM images available in Azure Marketplace that support `cloud-init`, and the list includes most of the prominent Linux distros. Documentation is available at `https://learn.microsoft.com/en-us/azure/virtual-machines/linux/using-cloud-init#canonical`.

- **Azure deployment script resource**: In ARM templates, we can bootstrap configurations such as managing user accounts, Kubernetes pods, or querying data from non-Azure resources. We use the `Microsoft.Resources/deploymentScripts` resource type to invoke the configuration. The deployment script uses a user-managed identity for authentication, and it requires an input to produce an output. Using the deployment script, we can execute PowerShell and Azure CLI commands and scripts. The progress of execution can be tracked from the Azure portal, Azure CLI, or Azure PowerShell. The deployment script supports advanced parameters such as execution environment, timeout options, and what to do with the resource if the script fails. For further details, refer to `https://learn.microsoft.com/en-us/azure/azure-resource-manager/templates/deployment-script-template`.

With that, we have completed bootstrap automation, let us look at configuration management.

## Configuration management

Bootstrap automation helps us with the post-deployment configuration; however, we cannot track the changes and ensure that the required packages are installed. Also, we need to ensure that they are compliant with the configuration we made. This is where the role of configuration management comes in. With the help of configuration management, we ensure that the packages needed are installed, and the desired state is achieved. Popular solutions for configuration management include the following:

- **Azure Automation State Configuration**: This solution supports VMs deployed on Azure, on-premises, and other cloud providers. Azure Automation State Configuration uses PowerShell **Desired State Configuration (DSC)** under the hood to maintain the configuration. In Azure **Automation Accounts**, we can upload a PowerShell DSC script, and we can assign the configuration to multiple resources from a centralized location. Each onboarded resource will be evaluated against the desired configuration, and changes will be made to align the current configuration to the uploaded configuration. Once the configuration is applied, the compliance status will be reported back to Azure, and we can confirm whether the configuration is the same for all resources. The following screenshot shows the **State configuration** dashboard and compliance:

Figure 4.3 – The Azure State Configuration dashboard

- **Chef**: An automation platform that can be used for infrastructure configuration management. Chef Clients needs to be installed on target machines that you want to manage using Chef. These onboarded machines can be monitored from one or more Chef servers, which is a VM with the Chef server role installed. Chef also offers additional components such as Chef Habitat and Chef InSpec for life cycle management and compliance requirements, respectively.

- **Puppet**: Application delivery and deployment process can be done with the help of Puppet. Similar to Chef, we need to install Puppet agents on the target machines, and these machines can be monitored from the **Puppet master**. We will use the Puppet master to execute the manifest files, which contain the configuration that we need to implement. If you are using DevOps, you can integrate Puppet with Jenkins and GitHub for developing workflows.

That is all we have for infrastructure configuration. Now we will learn how to automate operational tasks.

## Operational tasks

So far, we have discussed how we can automate infrastructure provisioning and configuration management. Once the deployment and configuration are done, we need to perform operational tasks such as resizing, patching, updating, and so on. Considering the scale of modern workloads, it is not easy to perform these tasks manually. This is where we need to automate operational tasks again, as this will eliminate human errors and improve productivity. Remember we discussed shutting down VMs during the weekend for cost optimization; this is also an example of an operational task that can be automated.

In Azure, we have multiple solutions that can be used for automation, out of which two popular ones are as follows:

- **Azure Automation**: You can write PowerShell and Python code to automate your operational tasks. There are thousands of ready-to-use templates available in the Azure runbook gallery, which can help you automate operational tasks without the need for coding. The following screenshot shows an excerpt from the Azure portal with a list of **runbooks**:

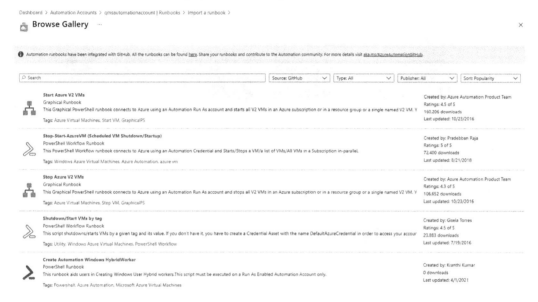

Figure 4.4 – The runbook gallery

You can also add custom PowerShell modules or Python libraries for developing your runbooks. These runbooks can be triggered using action groups, webhooks, schedules, or on demand. Automation accounts can also be used to manage non-Azure workloads.

- **Azure Functions**: This is a serverless computing solution that can be used for event-based automation. There are multiple triggers available for invoking **Function App**. We can develop functions directly from the Azure portal, as shown in the following screenshot, or use tools such as Visual Studio Code:

Figure 4.5 – Developing functions

The preceding code is a simple script written in PowerShell; you can write code in other languages such as C#, JavaScript, Java, Python, and TypeScript.

With that, we conclude automation; use automation in your environment to increase the deployment velocity, smoothen manual processes, eliminate human error, increase productivity, and boost employee satisfaction. Another key aspect of operational excellence is release engineering. Let us learn about the key characteristics of release management in the next section.

## Release engineering

Apart from automation, the optimization of **release management** is another goal we want to achieve using the operating excellence pillar. We have already seen the necessity of automation, goals, best practices, and types of automation in the previous section. Now we will discuss release management and different optimizations we can accommodate in our release pipeline. Release engineering is a lengthy topic; it is so vast that we could author a separate book on it. In this section, we will quickly cover the areas of optimization and key considerations. Links to relevant sections in the documentation will be provided for your reference. The stages we will cover are shown in the following diagram:

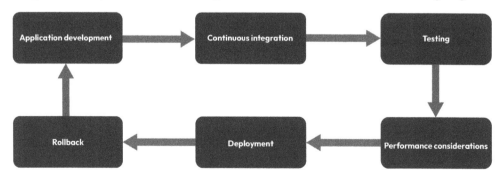

Figure 4.6 – Release engineering workflow

Let us take a look at the prime areas of release engineering.

## Application development

This is the first stage of release engineering, where developers are developing the application. In this stage, we need to ensure that modern application development strategies are adopted by the developers. The key factors that need to be considered in this stage are development environments and source control:

- **Development environment**: The environment in which you are developing the application should be equivalent to production so that during our emulation, we can understand the potential issues that we could face once we push the code to production. While setting up the environment as a replication of production, all dependencies should be installed so that we do not run into any compatibility issues when we move to production. Based on the solution that you are working on, the required dependencies will vary.

- **Source control**: This acts as a centralized repository for storing the code, and besides that, we can control, collaborate, and peer review code changes. Common examples of source control include GitHub, Azure DevOps, and BitBucket; all of these are powered by Git, which is a distributed version control system. Git also offers tools for developers for change tracking and collaboration. The basic process is to clone the existing code if there is any, suggest changes, and send the changes for peer review. If the change is approved, your suggestion will be merged with the existing code and resolve the conflict if there is any. With the help of version control, we can bring consistency and ensure that all developers are working on the same version of the code.

You can review the full documentation at `https://learn.microsoft.com/en-us/azure/architecture/framework/devops/release-engineering-app-dev`. After application development, the next stage in release engineering is **continuous integration (CI)**.

## Continuous integration

When we commit our changes to source control, this triggers a pull request. The pull request is reviewed and merged if approved; otherwise, it is rejected by your peers. When the pull request is created, the CI system is notified about this request. In other words, we are notifying the CI system that the code changes are ready to be integrated. With the help of CI, we can ensure that the code is ready for deployment, but this stage doesn't include the actual deployment of the code, which will happen at a later stage. When we set up continuous integration, we declare a set of baselines and tests; when we make the pull request, the code is evaluated against the baselines and tests we created. Feedback is provided to the developers based on the integration process conducted. If everything goes well, continuous integration produces and stages resources such as compiled code or container images if you are working on containerized applications.

Continuous integration ensures that the following actions are performed to help developers:

- Execute automated tests against code

- Ensure standards, quality, and configuration by executing code analysis

- Detect vulnerabilities by executing compliance and security checks

- Implement acceptance or functional tests to verify that the code operates as expected

- Detect issues and share feedback

- Produce deployable resources or packages such as compiled code, which includes updated code

In order to use CI, we need to use software solutions that support these integration pipelines. Popular choices are GitHub Actions (`https://docs.github.com/en/free-pro-team@latest/actions/guides/setting-up-continuous-integration-using-workflow-templates`) and Azure DevOps Pipelines (`https://learn.microsoft.com/en-us/azure/devops/pipelines/create-first-pipeline?preserve-view=true&view=azure-devops`).

With that, we move on to the next stage of release engineering: testing.

## Testing

While developing, testing is one of the elements that is inevitable. With the help of automation, we can achieve deployment velocity; however, we need to ensure that the code is assessed extensively before the deployment is done. A common misunderstanding is that we only need to evaluate the application code, but considering the best practices, we also need to evaluate our IaC and CaC templates. There is a wide variety of tools available for conducting different types of tests, examples are Azure Load Testing (`https://learn.microsoft.com/en-us/azure/load-testing/overview-what-is-azure-load-testing`), Azure Pipelines (`https://learn.microsoft.com/en-us/azure/devops/pipelines`) for automated tests, and Azure Test Plans (`https://learn.microsoft.com/en-us/azure/devops/test`) for performing manual tests.

Automated tests can be of the following types:

- **Unit testing**: Unit tests are performed as part of the CI runtime. You should treat unit testing as an extensive process and should include 100% of the code. The process should be streamlined and quick. Microsoft recommends keeping the test under 30 seconds as a baseline. Code syntax and functionality are verified in unit testing; if you are evaluating infrastructure templates, then we need to verify all resources described in the template are valid.

- **Smoke testing**: In this type of testing, we validate the code against a test environment that is equivalent to the production environment. With the help of smoke testing, we corroborate whether the system responds to the code and works as intended.

- **Integration testing**: In this test, we validate whether the application components can interact with each other as expected. Integration tests may take longer, depending on the number of application components.

Manual testing includes the following types:

- **Acceptance testing**: This test can be completed manually, but in certain scenarios, we can also automate the test. Partial or full automation can be done based on the context. In this test, we evaluate the compliance and system requirements to make sure that the code can be accepted.

- **Stress testing**: As the name implies, stress testing can be done to verify whether the application can withstand the changing load conditions. We monitor all components and identify the points where they hit bottlenecks. Baselines are established, and the tests are done against these baselines to confirm that the application can take the load.

- **Experimentation testing**: We need to deploy the code to the production environment and perform experiments. You can adopt popular experimentation strategies such as blue-green deployments, canary releases, and A/B testing for evaluating your application's functionality.

Furthermore, we have business continuity testing as part of the BCDR plan. Tests such as disaster recovery drills, exploratory testing, and fault injection are done to validate the business continuity and disaster recovery plan.

Once testing is done, next we have to consider the performance of our deployment infrastructure. Let us have a look at the performance considerations.

## Performance considerations

The performance of our deployment infrastructure is essential to achieving faster and more efficient builds. Poor performance means the build process will take longer, which will directly influence the deployment velocity. We can achieve builds by considering the following:

- **Opting for agents that meet your performance criteria**: The speed of the build process depends on the performance of your build machines. High-specification machines can decrease the build speed from hours to minutes. If you are using Azure Pipelines, then you have the ability to choose **self-hosted agents** and **Microsoft-hosted agents**. In self-hosted agents, you are able to manage the build machines, and you are responsible for the maintenance and upgrades of the machine. On the other hand, Microsoft-hosted agents are fully managed by Microsoft, including the updates, maintenance, and upgrades of the build server.

- **Location of the build server**: During the build process, code is fetched from the source repository and artifact repository, then it is processed by the build machine to deliver the compiled artifacts, test reports, code coverage results, and debug symbols. In this entire process, we need to confirm that the copy operations are quick. If the build machines are deployed in a different region than yours, then network latency and download speed will impact the overall process.

- **Build server scaling**: Based on the size of the product, we need to scale the build server. For a small product, a single server is enough. However, as the size and scope of the product expand, a single server may not be sufficient. The approach to managing these kinds of scenarios is to scale your infrastructure horizontally when you face performance bottlenecks. In Azure DevOps, you can use Azure DevOps agent pools to scale your build servers.

- **Parallel execution**: Having parallel build execution and testing can accelerate the pace of the build process and also reduce the time taken to evaluate the application.

The type of build you want to adopt depends on the size of the organization and your team's experience with the tools and processes. If you are able to achieve faster builds with the help of these considerations, we can proceed to the deployment phase.

## Deployment and rollback

In this phase, we will provision the required infrastructure with configuration and push our application code. As discussed multiple times in this chapter, we need to use automation for the deployment of resources and configuring them. You already know the rationale behind automation and why it is good. Once the deployment is done, document the process so that your team can learn and attain maturity. Let us take a look at the deployment considerations:

- **Automate infrastructure deployment**: Use Azure CLI, Azure PowerShell, or IaC (ARM templates, Bicep, Terraform, and so on) to provision the Azure resources.

- **Automate configuration management**: Consider the tools we discussed earlier, such as `cloud-init` (Linux), Azure Automation State Configuration, Puppet, and so on, for managing the configuration of your provisioned resources.

- **Automate maintenance tasks**: Operational tasks need to be automated as they are time-consuming and boring if you take the manual route. Azure Functions and Azure Automation Accounts are the best candidates for automating operational activities.

- **Deployment security measures**: With the help of Azure Policy, we need to ensure that we meet the organizational standards and compliance requirements.

- **Cutover process**: One process where we might face challenges with automation is the cutover from staging to production. If we do not oversee this properly, a cutover could lead to a production outage. Use deployment approaches such as blue-green deployment to have two production environments and minimize downtime.

- **Documentation**: Any processes involved in the procedure should be documented properly, including the failures. The team should use the failure as a learning opportunity and use the release documentation as their bible for deployment. Employees who created the process might leave the organization, and a new employee will take over. At that point, the process should be clear enough for the new employee to understand.

- **Staging workloads**: Never move your application directly to production; always stage your deployments. Staging helps us to perform tests and confirm the sanity of the application before moving to production. Updates to production are fully evaluated in a regulated manner and unforeseen deployment issues can be avoided with the help of staging.

- **Testing**: Always build your test environment equivalent to the production environment. If there is a divergence between these environments, unexpected issues may arise once we cut over to production. Once you are in production, it is not easy to diagnose and troubleshoot issues; better we resolve these problems in the testing environment before going ahead to production.

- **Logging and auditing**: Enable monitoring for your infrastructure and application to diagnose and troubleshoot issues. Azure Monitor offers infrastructure monitoring with the help of agents and Application Insights for monitoring your application.

- **High availability considerations**: From a high availability standpoint, consider deploying across multiple regions for global availability. If you are using a single region approach, plan how you will redeploy the application to a secondary region in case of primary region failure.

Even after all the testing, in some cases, you will face issues after deploying the code to production. Always anticipate deployment issues and plan how you can remediate them. Based on the solution that you are using, there are multiple ways to roll back your configuration to the last known best configuration. For example, if you are using the Standard plan or above of Azure App Service, then you can use deployment slots. You can have two versions of your code running under the same App Service Plan, say staging and production. After staging, you can swap your production and staging slots, which will promote your staging code as production, and the previous production code will be pushed to the staging slot. Now, if we face an unforeseen issue, we can easily perform another swap, and our previous production code will be back in action.

Along with deployment, you should also plan your rollback strategy. The process of rolling back depends on the tools and services your organization is using. After deploying the code, we need to monitor our application. Microsoft has shared a checklist for monitoring, which we will discuss in the next section.

## Monitoring

Modern applications have a lot of moving parts, and as the number of components increases, complexity also increases. With this complexity, we need to ensure that our system is monitored so that our components don't fail and the team is notified in the event of an issue. Azure Monitor is your go-to service for all your monitoring needs. We can collect data from disparate sources and analyze them in Azure. Furthermore, we can set up notifications (e-mail, phone, and text), automation (Azure Logic Apps, Azure Functions, Azure Automation runbooks, and Webhooks), and integration using **IT service management** (**ITSM**) connectors.

You need to ask the following questions within your organization to develop a monitoring strategy:

- Do you monitor your resources?

- Do you have application code monitoring using detailed instrumentation?

- Do you correlate application events across all application components?

- Do you analyze collected data to spot issues and patterns in application health?

- Do you have visualization tools to visualize monitoring data?

- Do you have alerts and notification plans to notify your team in case of an issue?

- Do you monitor Azure Service Health notifications?

- Do you monitor the application's health?

- Do you collect telemetry from the registered issues?

- Do you collect audit logs for compliance requirements?

Now take a moment and answer the preceding questions. If you have a strong monitoring strategy in place, then your answer should be positive for all the questions. If you missed something, have a look at the documentation and see why that question is relevant. You can refer to detailed information about these questions at `https://learn.microsoft.com/en-us/azure/architecture/framework/devops/checklist`.

Now we've understood the operational excellence pillar in depth. Once we take the assessment later in the book using a sample workload, you will have an even better understanding of this pillar.

## Summary

In this chapter, we covered the second pillar of the WAF, which is operational excellence. It has two foundational concepts: automation and release management. The entire chapter revolves around these two concepts. We started with the design principles that you should consider while designing applications aiming for operational excellence. Furthermore, we discussed automation, which is one of the prime areas of this chapter, as stated earlier. We covered the necessity, goals, and types of automation that we can use in our environment. Following that, we discussed release engineering, which is another key area of operational excellence. We discussed the end-to-end process beginning with application development, CI, testing, performance considerations, deployment, and finally, rollback. The last topic in this chapter was monitoring, which provides us with a checklist to verify whether we have a robust monitoring tactic.

Now that we are aligned on the operational excellence of our workloads, the next thing we need to ensure is that the applications are performing well. Azure offers different native solutions to improve application performance and delivery. In the next chapter, we will cover improving the performance efficiency of our workloads in Azure.

# 5

# Improving Applications with Performance Efficiency

**Performance efficiency** is the third pillar of the **Well-Architected Framework** (WAF). It refers to the ability of your solution to meet the business needs and demands placed on it by the consumers in an efficient way. In on-premises, we implement oversized servers or **virtual machines** (**VMs**) to manage unexpected load and demand. But in the cloud, this approach is not feasible; the main reason for that is the *cost*. As the chapter title says, we need to adopt an efficient approach to make sure that the performance of the workloads is not degraded. Before we plan to architect solutions in Azure, we should set aside the patterns we used to follow on-premises. One thing that hasn't changed compared to on-premises is you still need to forecast and anticipate a load increase in the cloud in order to meet the business requirements. Anticipating the load can help you optimize your workloads ahead of time and improve the user experience.

Leveraging the scalability of the cloud can improve efficiency in cloud computing. For example, if you are using an oversized VM in Azure to oversee the changing demands, it can manage that, but at the same time, the cost will be high. In other words, the potential of the VM is not fully utilized all the time; the full potential is used only when there is demand. Instead of this traditional approach, we need to adopt solutions such as **Virtual Machine Scale Set**, which can automatically increase or decrease the number of instances based on demand without increasing the cost, and is simultaneously performance oriented. Another advantage of the cloud is we can utilize the amazing telemetry to monitor the performance aspect of the workloads. Combined with the power of Azure Monitor, teams can immediately respond to incidents and resolve any performance-related issues.

As in previous chapters, we will discuss the various design aspects, followed by performance testing and monitoring. First, let's delve into the performance efficiency principles.

# Principles of performance efficiency

As we discussed in the introduction to this chapter, performance efficiency is the capability of a workload to become accustomed to adjustments based on demands made by consumers in an efficient manner. In order to improve the overall performance efficiency of the workload, we need to have a strategy. Microsoft has developed various **performance efficiency principles** to help you develop such a plan. The main principles are **design for scaling**, **testing**, and **monitoring**. Though we cover these principles in brief in this section, later in this chapter, we will take a deep dive into each of them to understand the underlying concepts of these principles. Let's take a look at the principles we need to leverage while aiming to attain performance efficiency.

## Principles for scaling

Elasticity is one of the advantages of the cloud, and **horizontal scaling** allows you to utilize this advantage. Based on the load, using horizontal scaling, we can increase or decrease the number of instances. These processes are also called **scale-out** (adding new instances) and **scale-in** (removing instances) of workloads. As discussed in *Chapter 3, Implementing Cost Optimization*, horizontal scaling helps to keep the cost down by removing the instances that are no longer required. In horizontal scaling, the following approaches are taken:

- **Describe a model that matches your business requirements**: With the help of load testing, evaluate the limits for when the load is predicted and when it is fluctuating. This is done to ensure that the application can scale. Once we know the scaling factor, we need to ensure that the service has enough quota and is supported in the region.

- **Use managed services**: PaaS solutions have built-in scaling configurations that can scale the workloads automatically to accommodate the load. Adopting managed services will remove the administration overhead and there won't be any chances of error.

- **Right sizing**: Finding the right size for the services is crucial while designing workloads. We need to perform load testing to understand the breaking point and decide on the size of the machine. Another factor is cost. We don't want to oversize the machine and create cost implications.

- **Adopt strategies during the design stage**: Leverage modern cloud capabilities, such as caching and autoscaling, during the initial stage itself so we don't need to make architectural changes later.

**Vertical scaling** is an alternative to horizontal scaling where we scale up or down. During the process, we add to or reduce the processing power of the machine. This requires a reboot of the machine as we are changing the hardware specifications. Furthermore, vertical scaling has an upper limit, and we will reach a point from which we cannot scale further. Once you reach the highest available size, the next option is to scale horizontally from that point. Due to this reason, it is better to utilize scale-out architecture in advance.

The next design principle is performance testing.

# Performance testing

It is always better to assess the performance of your application so that we can understand its limits and resolve issues at the earliest opportunity. Once you move to production, it is not recommended that you perform tests and make changes. It is better to perform the tests at the pre-production stage and make the changes accordingly so that we can confidently push our application to production. There are several approaches that you can take to evaluate the performance levels of your application, as described in the following list:

- **Execute performance load and stress test**: With the help of performance testing tools, we can measure the performance of our application by supplying programmed amounts of load to the application. The recommendation is to use data that is equivalent to production data so that we can estimate the performance of the application once it is moved to production. Furthermore, we need to determine the maximum load the application can withstand without service failure.

- **Establish baselines**: By calculating the current efficiency of the workload, we can set up a baseline that can be used to determine the maximum load our application can withstand. For example, if we are expecting 5,000 concurrent users to use your application, we can test with this value and make sure our application can deliver the content to the users without compromising performance. Based on our business growth plans, we need to document future plans to update the baseline. Say we expect an increase in the number of users to 7,000; then, we need to plan strategies ahead of time for these extra users we forecast.

- **Run test in the continuous integration (CI) pipeline**: In *Chapter 4, Achieving Operational Excellence*, we saw the importance of conducting tests in the CI build pipeline. To refresh your memory, we want to detect issues early on in the build process itself so that remediations can be applied before we move to production. The code and queries will be evaluated to ensure that they are not causing an adverse effect on the performance of the application.

We will cover the performance testing checklist and the tools later in this chapter so that you get a better understanding of the process. With that, we will now briefly cover the third and last performance principle: monitoring.

# Performance monitoring

Even though we conduct tests to attain maximum performance of the application, there will be still exceptions due to transient errors, service degradation, and hardware failures. We need to ensure that the performance of the application is continuously monitored to capture performance bottlenecks. Usually, we don't monitor resources when they are in the development phase, and monitoring is only required once the workloads are in production. Having said that, it is better to monitor the workloads in the development phase as well because we will be able to capture bottlenecks in the early stage itself and we don't have to make changes when they are in production. The following approaches should be taken into consideration for monitoring the performance of your application:

- **Monitor the health of the entire solution**: Your application will have multiple components and dependencies; we need to monitor each of these components to ensure the optimum performance of the application is attained.

- **Capture data via an iterative process**: Assess performance metrics over time to understand the historical usage of the application. With this data, we can understand the performance bottlenecks and usage patterns, which can be further used to develop autoscaling strategies.

- **Reassess workload requirements continuously**: Remember that there is always room for improvement. Business requirements may change over time, and we need to reassess our workloads to adopt these recent changes. These processes include changing obsolete configurations and replacing deprecated components.

Now that you are familiar with the pillars of performance efficiency, let us discuss the checklists for each of these principles. We will begin with the designing for performance efficiency checklist.

# Design for performance efficiency checklist

In the previous section, we covered a high-level overview of the principles of performance efficiency. Now, we will take a deep dive into each principle, starting with designing for performance efficiency.

## Application design

We need to design the application with performance in mind. If we don't, then our application cannot perform well and will affect the user experience, further affecting the business. In order to manage the load, we need to ensure that the design is appropriate. The following are the application design considerations for applications:

- **Design for scaling**: We can react to the application load with the help of scaling. In scaling, we will be able to increase or decrease the number of instances. All applications should be designed with scalability in mind. We covered the importance of scaling earlier in this chapter when we discussed principles of performance efficiency.

- **Plan for growth**: We need to plan additional resources to accommodate business growth. Depending on the service limits of your Azure subscription, you will have quotas associated with each service. Once you reach the upper limit, you need to have plans to expand in advance. This will make the scale-out process easy and streamlined. In this way, resources won't be lacking when they are required. For example, the maximum number of ExpressRoute circuits per region per subscription is four. If you need more circuits in the same region, then you need to make plans for that.

- **Use platform autoscaling features**: In Azure, we have multiple solutions, and some of them support scaling features. It is recommended that we use the native autoscaling features rather than using homegrown autoscaling solutions. Since we are using native metrics to trigger autoscaling, the scaling will happen quickly as soon as the threshold is met. Third-party solutions may demonstrate a delay in the scaling as they need to collect the metrics from Azure using an agent and then initiate the scaling process.

- **Decouple workloads**: Rather than adopting monolithic architecture, try partitioning the solution into small components so that you can oversee the performance of these components independently. Remember, we discussed microservices architecture in *Chapter 4, Achieving Operational Excellence*, which is an example of this approach.

- **Prevent client affinity**: Ensure your application doesn't require client affinity wherever possible. By preventing client affinity, the requests can be routed to any of the backend instances without the need to manage the state information of each user session. In Azure, we have multiple load balancers that support client affinity and session persistence; however, the recommendation is to ignore this element wherever viable.

- **Offloading intensive tasks**: Performance may get degraded if you have long-running tasks. These tasks will consume a significant amount of resources, and this is totally anticipated. The way we can avoid this performance degradation is by offloading these intensive tasks to an independent task. Based on the hosting platform that you are using, there are multiple ways to offload this, such as setting up worker roles or background jobs. In this way, the intensive tasks are taken care of separately and will not lead to a bottleneck.

- **Multiple compute nodes for background tasks**: In the previous point, we covered the need to offload intensive CPU or I/O tasks as background tasks to avoid performance degradation. While offloading, consider using multiple compute nodes to finish the task quickly. The rationale behind offloading is they are CPU or I/O intensive, so by splitting the load over multiple compute nodes, we can process the tasks and deliver results quickly.

With that, we have completed the application design considerations. Next, we will cover the data management aspects for improving the performance of our applications.

## Data management

While designing solutions with performance in mind, we tend to believe that it is all about setting up a beefy infrastructure. Let's say you have an application that retrieves data from a database and displays it on a web page. This also includes other data such as images and documents. We have already set up the infrastructure and ensured we have the right sizing; however, if the database queries are not properly optimized, that will lead to latency in data retrieval. So, it's important that we consider enhancing the performance of our data solutions to make our application optimal. The following are the data management considerations that should be considered while designing applications to improve performance:

- **Data partitioning**: Consider dividing your data across multiple databases and storing them in different database servers. Using this approach, we can elevate the performance and allow simpler scaling. Multiple partitioning techniques exist, such as horizontal, vertical, and functional. In order to improve query performance and ensure easy scaling, flexible management, and higher availability, we can use a combination of these partitioning techniques.

- **Consistency**: If your data is read frequently and not written often, then **eventual consistency** is the right data consistency level for you. Eventual consistency will improve the scalability by lowering or eliminating the time required to orchestrate data partitions across multiple database stores.

- **Decrease chatty interaction between application components**: If your application is required to make multiple calls to a service, that is not an ideal solution. Multiple calls would lead to latency in response and would end up in performance degradation. Typically, multiple calls should be combined, and a single call should be made. Optimization of complex queries and easy performance monitoring can be achieved by reducing these kinds of chatty conversations.

- **Use queues**: Too many requests to a service can cause service failure if the service is not scaled enough. These spikes in usage can be managed using queue-based load balancing. Requests will be sent to the queue and they will be processed one by one to avoid the service crashing.

- **Avoid data processing load on the data source**: Any data processing, such as of XML or JSON objects, should be done within the application. If you do this processing in the datastore, that will be an extra load on the datastore, causing it to fail. Instead of processing on the datastore, let your application serialize the data, transform it based on the business requirements, and send it to the database.

- **Volume of data retrieved**: Optimize your queries and retrieve only the required rows. Querying and retrieving a substantial number of rows can take longer and will affect the performance. The best practice is to avoid retrieving data unnecessarily by using table value parameters and appropriate isolation levels.

- **Use caching**: If your data is getting accessed frequently, then you should ideally use caching. Using a caching solution, we will cache the data and applications will read the data from the cache instead of directly querying the database. Older data will be removed from the cache. If an application queries the data, then that will be sent to the datastore, and the results will be cached. Any subsequent query responses will be taken from the cache.

- **Data retention**: Over time, the database size will grow. As the data grows, the storage cost will increase, and queries may take longer as they need to look at a larger dataset. It's recommended that you archive older data (move it to long-term storage) when no longer required or accessed to optimize the cost and performance.

- **Client-side caching**: By default, web applications do not cache content. It's advised that you enable cache settings on the server to cache the contents that can be cached, which can improve the performance.

- **Use Azure Blob Storage and Azure Content Delivery Network (CDN)**: With the help of these services, we can reduce the load on the application. For example, if your application is storing static public content such as banners, images, and documents, then this can be stored in Blob Storage. The application can dynamically pull this data from the storage and display it. Furthermore, we can leverage Azure CDN to accelerate content delivery. Azure CDN can be integrated with Azure Blob Storage, allowing it to cache the content to edge locations. Content stored in the CDN cache will be delivered from the nearest geographical edge location to the users.

- **Optimize and fine-tune SQL queries**: A number of T-SQL queries may have unfavorable effects on the performance of the database. This can be reduced by optimizing the query and storing the queries as a stored procedure. For example, if you are storing the date and time as a string in the database, you need to delimit the string and manipulate the information. A better approach would be storing the date time as a `datetime` object so that you can leverage datetime functions for comparison and easy manipulation.

- **Denormalize data**: Data duplication and inconsistencies can be avoided by denormalizing the data. As mentioned earlier, any data transformation should be done by the application itself to avoid load on the database.

The aforementioned considerations should be made while designing solutions with performance in mind. Next, we will learn about the implementation principles for performance.

## Implementation principles

Performance considerations should be adopted during the implementation stage as well. Microsoft recommends that we follow these best practices while designing solutions:

- **Avoid performance antipatterns**: In the cloud, we have multiple cloud design patterns (https://learn.microsoft.com/en-us/azure/architecture/patterns/); however, there are different design patterns that could cause a negative effect and scalability issues when the application is under load. Review performance antipatterns (https://learn.microsoft.com/en-us/azure/architecture/antipatterns/) while designing applications to avoid this issue.

- **Use asynchronous calls**: When you are calling services that have I/O or network bandwidth limitations, consider using asynchronous calls. By following this approach wherever possible, the calling thread will not be locked.

- **Avoid locking access to resources**: Develop optimistic approaches to accessing resources instead of locking access. For example, if your application needs to authenticate every time before writing to the storage account, then it will impact the performance. You can use a managed identity and store the account key in Azure Key Vault to improve performance rather than going for other modes of authentication.

- **Use HTTP compression**: If your network has low latency, then it's better to use HTTP compression in your application. When we access the application, most of the content generated by the application is sent over the network using HTTP. If the amount of data is high, then you will face performance issues, and HTTP compression is the solution for this.

- **Minimize the connection time**: Retain connections and resources only for as long as you need them. Having a longer connection time will make the application provide more resources. While designing the solution, make sure you set an appropriate connection time so that once the connection time is over, the resources are released.

- **Minimize the number of connections**: The higher the number of allowed connections, the more compute power we need. We need to restrict the number of connections and also reuse open connections whenever feasible; this can enhance the performance of the application.

- **Optimize network use and send requests in batches**: In the *Data management* section, we discussed how we can use queues to optimize the number of requests. While reading from the queue, read or write the requests in batches while accessing storage or cache. This pattern can improve network use and boost the efficiency of the services.

- **Avoid storing server-side session state**: Remember, we covered the need to prevent client affinity in the *Application design* section. Server-side session state management requires client affinity, and this is something we need to avoid. The recommendation is to develop an application as *stateless* if you don't need to manage the state. However, if you need to have state management, then you need to incorporate caches to store this information in a centralized location rather than storing it independently.

- **Optimize the table schema**: While querying data stored in Azure Table storage, often we need to pass the table name and column name to the query. Consider using short names to avoid this overhead while keeping the readability and manageability of the columns.

- **Configure dependencies on the fly**: Instead of making multiple calls to check whether a resource exists before creating it, consider using out-of-the-box methods that can check the existence of the resource and create it if it's not present in a single call. For example, you can use the `CreateIfNotExists` method in the Azure Storage SDK to create a table if it is not present.

- **Use lightweight frameworks**: Use APIs and frameworks that can minimize resource usage, improve execution time, and diminish the overall load on the application.

- **Minimize the number of service accounts**: The more service accounts accessing the resources or services, the more connections will be made to the application. As mentioned earlier, we need to reduce the number of connections, and the best way to do that is to reduce the number of service accounts accessing the application.

With that, we have finished exploring the design principles. Next, we will look at how we can assess our application for performance, scalability, and resilience.

# Testing for performance efficiency checklist

Before we publish our application, we need to assess and validate whether it can manage the expected usage. Besides that, the application should be configured to manage unexpected load using scaling principles. Using the tests, we will be able to capture issues and resolve them before we move the application to production.

With the help of performance testing, we can ensure our applications are efficient, responsive, and scalable and the performance of the application meets the business needs. The diagnostic data that we collect as part of the tests can be used to confirm whether our application can withstand the load, and this is essential to identify bottlenecks. As you know, bottlenecks happen when there is a disruption in the data flow due to insufficient capacity to oversee the load.

We have **performance testing**, **load testing**, and **stress testing**. We can consider performance testing as a superset of load and stress testing. The benchmark behavior of the application can be calculated using performance testing; in fact, this is the primary goal. Load testing, as the name implies, performs tests on the application by progressively increasing the load until it reaches a point where it breaks. Furthermore, stress testing involves various tasks that will overload the application to test its limits and boundaries of operation. Let's start with considerations related to performance testing.

## Performance testing

The rationale behind **performance testing** is to identify problems and fix them before they reach the end users. One of the best practices associated with performance testing is to plan a **load buffer** to adapt to unplanned spikes without overburdening the infrastructure. For example, if your application is expecting a normal load of 150,000 requests, then you need to plan your infrastructure to manage 150,000 requests at 80% of the infrastructure capacity. In this way, you are adding a buffer so that even if the requests increase to 160,000, your infrastructure should be able to manage the spike. If you anticipate an increase in the load due to any event and the current infrastructure cannot manage the spike, then you need to upgrade the size (SKU) of the infrastructure. If you are planning geo-redundancy, then you have to make sure the selected SKUs are available in the secondary region as well.

When it comes to performance testing, we need to review our application architecture in light of the following considerations:

- **Collaboration is important**: As mentioned earlier, performance testing is mandatory for all applications to achieve performance efficiency. The testing cannot be done by developers alone. They could perform tests on the developer workstation; however, we cannot consider this full-fledged testing. The right approach is to conduct these tests in a test environment that has a configuration equivalent to the production environment. To build this test environment, we need dedication from not only developers but also architects, database administrators, platform administrators, and network administrators. In short, we need cross-team collaboration to make the testing successful.

- **Capacity planning**: Based on the type of application, you need to anticipate the fluctuation in load and plan the capacity. For example, if your application is used for tax filing, you need to anticipate more users toward the last date for filing taxes. If you don't consider this load, then your application will collapse and none of the users will be able to file their taxes. Application load can be impacted by several events, such as political, economic, weather changes, holidays, and deadlines. Moreover, you need to make certain that your subscription has enough quota in the region to manage the scaling.

- **Leverage existing tests or create new tests**: We have multiple types of testing within performance testing, such as load testing, stress testing, API testing, and client-side testing. While performing these tests, teams should be aware of the pros and cons of each type of testing.

- **Testing at all stages**: In testing, we are not just evaluating the application code. We are evaluating the infrastructure automation and fault tolerance as well. This will help us in evaluating whether the application is working and confirm that it works as expected in a development, pre-production, and production environment. Since we are integrating testing from the initial stages of the application, we can easily capture errors at the earliest opportunity and fix them before we ship the application to production.

- **Plan infrastructure using test results**: Performance testing has two subsets: load testing and stress testing. With the help of these tests, we can determine the upper limit and maximum point of failure of the application. Based on the results, we can plan the infrastructure capacity to manage the spike.

- **Multi-region failover testing**: Using the help of a paired region in Azure, we can easily scale our application across regions. If the primary region goes offline, we can failover to a secondary region. By performing regional failover, we can understand the time taken to recover and route requests.

As you already know, performance testing can be of two types: load testing and stress testing. We will start with load testing.

## Load testing

In **load testing**, we assess the system performance as the demand (or number of requests) increases. Using this test, we can identify the point at which the application breaks. This test is conducted with normal and heavy loads to understand the reaction of the application so that we can fix issues before we ship the application to production.

Over a given interval of time, we will use virtual users or simulated requests to measure the load. With the help of load testing, we can build insights into how and when our application needs to scale. Furthermore, we need to ensure that the SLA of the application is met. The latency between application components and microservices can be easily calculated with the help of load testing.

The following key points should be considered while performing load testing:

- **Know the Azure limits**: As you know, Azure services come with limits. These limits are imposed on a subscription, or sometimes at the regional level as well. Some limits are hard limits, which you cannot change, while others are soft limits, which you can modify by reaching out to Microsoft Support. You can review the limits at `https://learn.microsoft.com/en-us/azure/azure-resource-manager/management/azure-subscription-service-limits`.

- **Calculate typical loads**: Measuring the typical and maximum load on the system helps you understand the capacity of your application. We need to implement monitoring solutions to understand the application usage patterns.

- **Test under several scales**: Initially, evaluate your application with a typical load and gradually increase the load to understand the application behavior. If you have implemented autoscaling, then verify whether the scaling works as expected when the threshold is reached.

Now that you are familiar with the concepts of load testing, we will cover stress testing.

### Stress testing

While load testing concentrates on evaluating what the system can manage, **stress testing** focuses on estimating the breaking point by overloading the application. System stability is measured along with its capability to tolerate extreme increases in load. As the name implies, we are putting stress on the application to find the breaking point. This testing involves sending the maximum number of requests to the application from a service that the system can manage at a given time without compromising the performance of the application. This telemetry can be used to understand the upper limits of your application and will provide insights for shaping your infrastructure to accommodate the requests.

We will take a T-shirt size approach here, by using the infrastructure that matches your application metrics. With the preferred size, put stress on the workload to determine the maximum load supported and compute the operational margin. Once you calculate the acceptable operational margin and response time under a typical load, check whether the current infrastructure can manage this without any performance degradation. If you see any performance issues, then reshape your infrastructure adequately. You can oversize the instance and avoid performance degradation; however, oversized instances come with a price. If we oversize, then we will have cost implications; if we undersize, then we will face deficient performance. So, it's very important to find the right balance between cost and performance.

Stress testing is not all about increasing load. We can perform additional tests, such as reducing resources and increasing latency. These tests are crucial in determining the worst-case scenario and application behavior.

Though performance testing is primarily of two types, namely, stress testing and load testing, we can also perform multi-region testing if we are planning to deploy our workloads at a global scale for high availability. So, before we conclude this section, let's quickly cover the **multi-region testing** of applications that require global redundancy.

### *Multi-region testing*

If you are deploying to a single region, you can leverage high-availability techniques such as **availability sets** and **availability zones**. One limitation is that these techniques are scoped to a single region, and if this region goes offline, your application will be unavailable. However, with the help of load-balancing solutions such as Azure Front Door, we can rapidly expand our infrastructure to a global scale. In such a scenario, even if the primary region goes offline, we can failover to a secondary region.

One of the reference architectures for multi-region deployment is available at `https://learn. microsoft.com/en-us/azure/architecture/reference-architectures/ app-service-web-app/multi-region`.

According to this architecture, the same application is running in two regions (active and standby) load balanced using Azure Front Door. If you are following this type of architecture, you need to find the time taken to reroute the request to the secondary region in case of an active region failure. There are different routing methods available in Azure Front Door to manage the routing of requests to the backend. You can review this at `https://learn.microsoft.com/en-us/azure/ frontdoor/front-door-routing-methods#priority-based-traffic-routing`. Ideally, you should conduct failover drills to make sure the application is working as expected in the secondary region same as it would work in the primary region.

So far, we have covered the theoretical aspect of how the testing is done. Now, we will take a look at considerations about the testing tools available for us to conduct the previously mentioned tests.

## Testing tools

There are a plethora of tools available on the market that can be utilized at various stages of the application development process. Azure Load Testing (`https://learn.microsoft.com/ en-us/azure/load-testing/overview-what-is-azure-load-testing`) is an example of such a tool that is natively available in Azure to simulate load and different usage patterns. The following key points should be considered when deciding on testing tools:

- **Choose tools based on the type of performance testing**: As mentioned earlier, depending on the application framework, there are different tools available to evaluate the capability of the application. Some examples are JMeter, Selenium, k6, and Azure Load Testing. Each of these tools targets distinct aspects of testing. For instance, to test against endpoints and validate the HTTP status, we can use JMeter. On the other hand, if we want to check data quality and variations, then we can use Selenium or k6. In Azure Load Testing, we can include an existing JMeter script and identify performance issues by monitoring the application. Furthermore, we have **application performance monitoring** (**APM**) tools such as Application Insights to monitor application performance; however, this tool is not designed to evaluate server load.

- **Perform performance profiling and load testing**: During the application development process, we need to evaluate every component before we ship the final product. All testing should be done on the same hardware as the production platform with the same data type, user load, and quantity of data to understand the application behavior.

- **Decide whether it is better to use automated or manual testing**: As you are already aware, we can perform automated and manual testing. **Automated testing** is best as it can be performed without human intervention. Based on the number of tests that you are conducting, you can divide them between automated and manual testing. **Manual testing** should be limited and shouldn't be performed frequently as it requires a lot of time for completion.

- **Caching to enhance performance**: Caching is more beneficial depending on the amount of data you have. If you have immutable data, or data that doesn't change frequently, then caching is the best solution.

- **Decide how you will manage static content in debug and release**: While you are evaluating the application in debug mode, you can load the static content from local storage. However, when we evaluate this in release mode, all the static content should be loaded from the preferred CDN.

- **Gauge the application performance of each workload**: Try simulating different workloads on your application and gauge the performance of each workload. In doing this, we will have a clear understanding of the resources needed to host our application. With the help of performance indicators, we can conclude whether the application is working as expected or not.

With that, we conclude our discussion on testing tools. Now, we will learn how to monitor the performance of applications.

## Monitoring

Performance issues can result from database queries, latency between services, memory leaks in code, or under-provisioned infrastructure. Without telemetry, it will be extremely hard to troubleshoot and debug these kinds of issues. This is where we need to implement monitoring to fill the gaps and collect telemetry. Monitoring solutions should be designed to monitor the following factors:

- Scalability
- The resiliency of the application, infrastructure, and dependent services
- The performance of the application and infrastructure

Therefore, the following checklist should be used while designing monitoring solutions:

- In order to get full insights into your application, you need to enable collection on your Azure resources. Once configured, you will be able to ingest and store the telemetry to build visualizations.

- In Azure, some metrics are readily available out of the box without any additional configuration. These metrics include CPU, memory utilization, network in/out, and storage utilization. We need to focus on monitoring these metrics.

- Resource and platform logs should be ingested into a Log Analytics workspace to capture events and incidents.

- Monitor the usage patterns of your application. This will help you in building scaling rules.

- Once you configure the Log Analytics workspace to ingest logs and metrics, write custom queries to find abnormalities. Furthermore, we can set up alerts to notify administrators about these abnormalities based on the query response.

- By default, the Azure Log Analytics workspace offers 31 days of free data retention; after that, older logs will be unavailable. It's recommended that you increase the retention period so that you have a larger time period than just a month. You need to pay extra for the extra retention period that you opted for.

Now we will look at different levels from which you can collect monitoring data in Azure. All the monitoring is done from the Azure Monitor service but there are different services corresponding to each level.

## Azure Monitor

**Azure Monitor** is a comprehensive monitoring solution available in Azure that can be used to collect, analyze, visualize, and respond to telemetry from resources. With the help of Azure Monitor, we can pull telemetry from Azure and non-Azure environments, such as on-premises, Amazon Web Services, and Google Cloud Platform. The Microsoft documentation provides a pictorial representation of Azure Monitor that shows its capabilities. The following figure shows a high-level overview of Azure Monitor (`https://learn.microsoft.com/en-us/azure/azure-monitor/overview#high-level-architecture`):

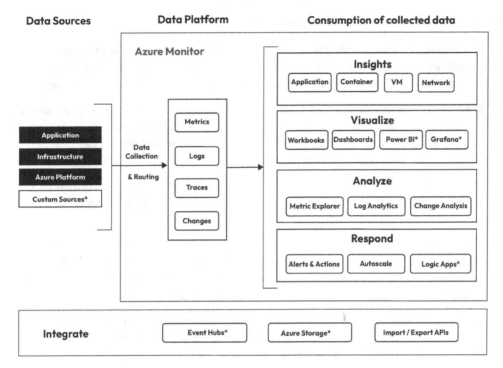

Figure 5.1 – High-level overview of Azure Monitor

As shown in the figure, we can collect data from discrete data sources, and then that can be used to build insights and visualization and analyze and respond to telemetry. The collected data includes metrics, logs, traces, and changes that are available in the data source. Azure Monitor is a vast topic. It is so vast that we could write a separate book on it. For now, we will limit the scope to topics related to performance monitoring. When it comes to performance monitoring, we have two major types: application monitoring and infrastructure monitoring. We will start with application monitoring.

## Application monitoring

We have different APM solutions available on the market. Application Insights is the solution provided by Azure Monitor to monitor your application. We can integrate Application Insights with development and production applications as well. By doing so, we can take a proactive or reactive approach. With a proactive approach, we can understand how the application is performing, and with a reactive approach, we can confirm the cause of an issue by analyzing the application execution data. Apart from the metrics and telemetry that Application Insights can collect, it can also help collect and store trace logging data.

Application Insights provides the following features:

- **Live metrics**: Without affecting the performance of the hosting environment, we can examine live activity from our application. We will be able to see metrics and other application logs in real time. This can be quite useful if you are debugging or trying to reproduce an issue.

- **Availability**: We can investigate our application endpoints to calculate the overall availability and responsiveness of the application. This process is also called **synthetic transaction monitoring**. You can set up multiple availability tests for your application and queries can be sent from multiple regions to calculate the latency from discrete locations.

- **DevOps integration**: Based on data collected by Application Insights, we can create work items in Azure DevOps or GitHub. These work items can be used by developers to fix bugs, prioritize feature requests, and so on.

- **Usage pattern**: Understand which features are popular with users and how the users are interacting with your application using Application Insights. These patterns can help in improving the user experience. If fewer users are using a feature, we will get to know that from the usage pattern. Furthermore, if a new feature is added, we can check how users are reacting to it.

- **Smart detection**: Using proactive telemetry analysis, we can detect failures and anomalies automatically. With the help of Azure Monitor alerts, administrators can be notified if there is a sudden spike in failure rates or any deviation from the regular usage pattern.

- **Distributed tracing**: For a given execution or transaction, we can search and visualize the flow of data from end to end. This will help us to correlate logs if we are troubleshooting microservices where the application components are decoupled.

- **Application map**: This will provide a high-level view of the architecture of the application components and dependencies along with the ability to review the health and responsiveness. In this logical diagram of your application, each component will be represented by its role or name. In the case of a distributed application, we can easily spot performance bottlenecks from the application map.

Onboarding to Application Insights can be done in two ways, using an agent (**auto-instrumentation**) or using the Application Insights SDK in your code. Many frameworks are supported by Application Insights, such as C# | VB (.NET), Java, JavaScript, Node.js, and Python, and the solution can be used with applications running in both Azure and non-Azure environments. Auto-instrumentation requires zero code changes and is supported for ASP.NET (hosted with IIS), ASP.NET Core (hosted with IIS), and Java. Other frameworks will require code changes to be made using the SDK for telemetry collection regardless of where they are hosted.

We will now take a look at some ready-to-use dashboards available in Application Insights to monitor your application. The following screenshot shows the failures that occurred in the application:

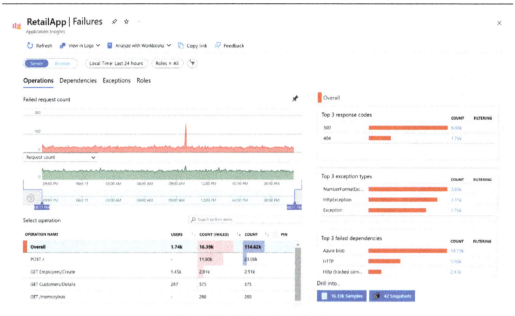

Figure 5.2 – Failures in application

In the figure, we can see the number of requests that came to each path and the number of failures for the respective path. Furthermore, we can see the major exception types, error codes, failed dependencies, and so on. Similarly, the following figure shows an availability dashboard where we can set up tests to monitor the health of the endpoint:

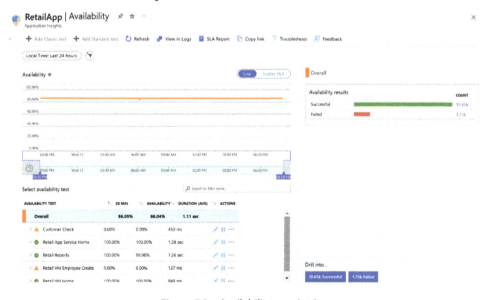

Figure 5.3 – Availability monitoring

As shown in the preceding screenshot, we can see the overall availability of the application and the availability of individual endpoints.

In the following screenshot, we can see the performance of the application and the average duration for each request:

Figure 5.4 – Performance monitoring

As described in the features of Application Insights, we have other dashboards, such as Live Metrics, Smart Detection, and Transaction Search. With that, we will move on to infrastructure monitoring.

## Infrastructure monitoring

In the previous section, we saw how we can monitor application-level telemetry. Along with the application, we need to monitor the infrastructure. If infrastructure failure happens, your application will not be able to work even if the issue is not related to the application. Infrastructure refers to the platform where we are hosting our application. Luckily, Microsoft provides an out-of-the-box solution to review some of these metrics; advanced monitoring requires additional configuration.

Primarily, we need to focus on the metrics and logs collected from the platform. If you recall the high-level architecture of Azure Monitor (refer to *Figure 5.1*), we have the metrics and logs datastores. These datastores serve as the primary source for monitoring our infrastructure. The data stored in these datastores can be further ingested into Azure Log Analytics, where we can run queries and analyze the data. Let's discuss platform metrics first.

## Platform metrics

If you are using a Windows machine, then you are most likely aware of **Task Manager**. In Task Manager, we can see various performance counters, such as CPU, memory, network, and disk IOPS. If the values for these counters are high, then that means there are performance bottlenecks. When it comes to Azure, we should be able to monitor these kinds of metrics to understand the state of the system. Basically, **metrics** are numerical values that describe the state of the system plotted in a time series. Each service in Azure exposes a set of metrics that can be used for performance monitoring. The availability of metrics will vary from service to service. For example, if you look at VM metrics, you will have CPU, memory, network, and so on. However, if you check Azure Storage, then the metrics will be transactions, ingress, latency, and so on. In order to view these metrics, you don't have to make any configuration changes to your resources; metrics are collected the moment the resource is deployed.

For instance, once a VM is created in Azure, we can navigate to **Overview | Monitoring** and see the platform metrics, as shown in the following figure:

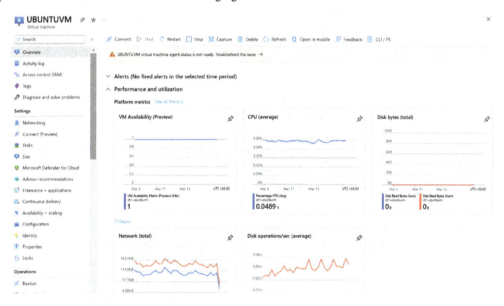

Figure 5.5 – Reviewing platform metrics

Don't assume that these are the only metrics available for the resource. If you look at the **Metrics** blade, Azure offers a plethora of metrics that can be quite useful in building your monitoring strategy:

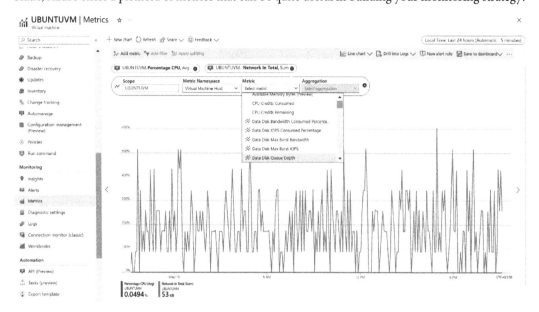

Figure 5.6 – Available metrics

As you can see in the preceding figure, we can review multiple metrics from multiple hosts on a single screen. In the preceding example, we are checking the **Percentage CPU** and **Network In Total** metrics for a VM. Furthermore, we can also see a list of available metrics that can be added to the graph for analysis.

The data that we are seeing in **Metrics** is basically **platform metrics**. In other words, these metrics are collected from the host machine on which your VM is deployed. We also have the capability to get guest-level metrics, but this requires additional configuration. Also, this data needs to be stored as we will not analyze it from the portal like platform metrics. Ideally, we would enable **Diagnostic Settings** to collect the data and then send it to datastores such as Log Analytics (preferred) or Azure Storage. Since this requires storage, you will be billed for this; however, platform metrics are completely free. For example, if we take Azure Web Apps, we have platform metrics available just as we saw in the case of VMs. Additionally, we can configure diagnostic settings to collect metrics at the guest OS level and send them to different destinations, as shown in the following screenshot:

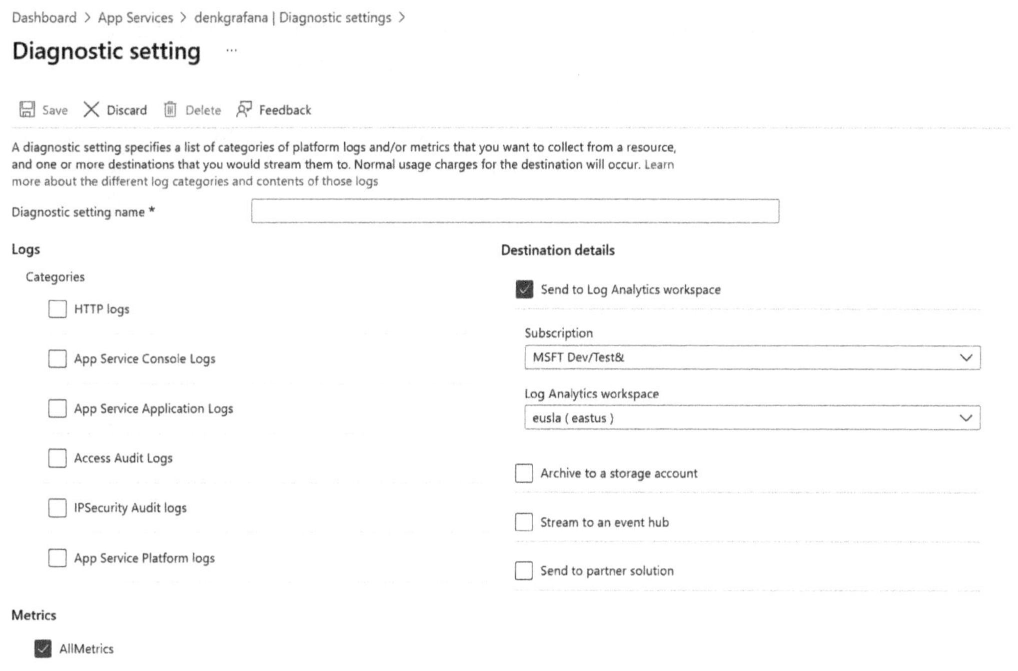

Dashboard > App Services > denkgrafana | Diagnostic settings >

## Diagnostic setting    ···

🖫 Save    ✕ Discard    🗑 Delete    ⧉ Feedback

A diagnostic setting specifies a list of categories of platform logs and/or metrics that you want to collect from a resource, and one or more destinations that you would stream them to. Normal usage charges for the destination will occur. Learn more about the different log categories and contents of those logs

Diagnostic setting name *      [                                        ]

**Logs**                                          **Destination details**

Categories                                        ☑ Send to Log Analytics workspace

☐ HTTP logs                                       Subscription
                                                  [ MSFT Dev/Test&                    ⌄ ]
☐ App Service Console Logs
                                                  Log Analytics workspace
☐ App Service Application Logs                    [ eusla ( eastus )                  ⌄ ]

☐ Access Audit Logs                               ☐ Archive to a storage account

☐ IPSecurity Audit logs                           ☐ Stream to an event hub

☐ App Service Platform logs                       ☐ Send to partner solution

**Metrics**

☑ AllMetrics

Figure 5.7 – Enabling diagnostic settings

In the preceding figure, we are collecting all metrics at the guest OS level and sending them to the Log Analytics workspace. We will have a similar option available for each resource in Azure, where we can enable diagnostic settings and ingest data for analysis. Now, we will move on to the next topic – platform logs.

## Platform logs

In Azure, we can collect operational logs at the platform and resource level as we have seen in the case of metrics. As the name implies, the logs are usually structured, which provides insights into what events happened, what changes were made, and more. Logs are quite useful in tracking changes and errors. The structure of the logs generated will vary from service to service and solution to solution. Logs will be ingested into a Log Analytics workspace and we can run complex queries to analyze the data. With Azure Monitor logs, we can collect logs in near real time, making it an ideal candidate for capturing issues and alerting administrators.

At the platform level, we have Azure activity logs, which provide us with insights into different ARM operations done on our subscription. For example, we can see at what time a VM was restarted, who restarted it, and the status of the operation, as in whether it failed or succeeded. By default, Azure provides 90 days of retention for activity logs. If you need to store data for more than 90 days, you can send the logs to a Log Analytics workspace and define the retention policy at the workspace level. In the following figure, you can see the **Activity log** at a subscription level, which shows events, timestamps, details of the user who initiated the process, and so on:

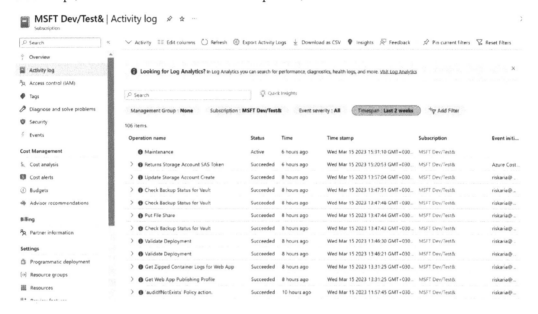

Figure 5.8 – Azure Activity Logs

In the preceding screenshot, you can see the **Export Activity Logs** option at the top, which can be used to send the logs to a Log Analytics workspace. As we have seen in the case of metrics, for the complete monitoring of performance, we need insights into the log generated by a guest OS. In *Figure 5.7*, we enabled only metrics for the resource; however, as shown in the following screenshot, we can also send the logs to different destinations:

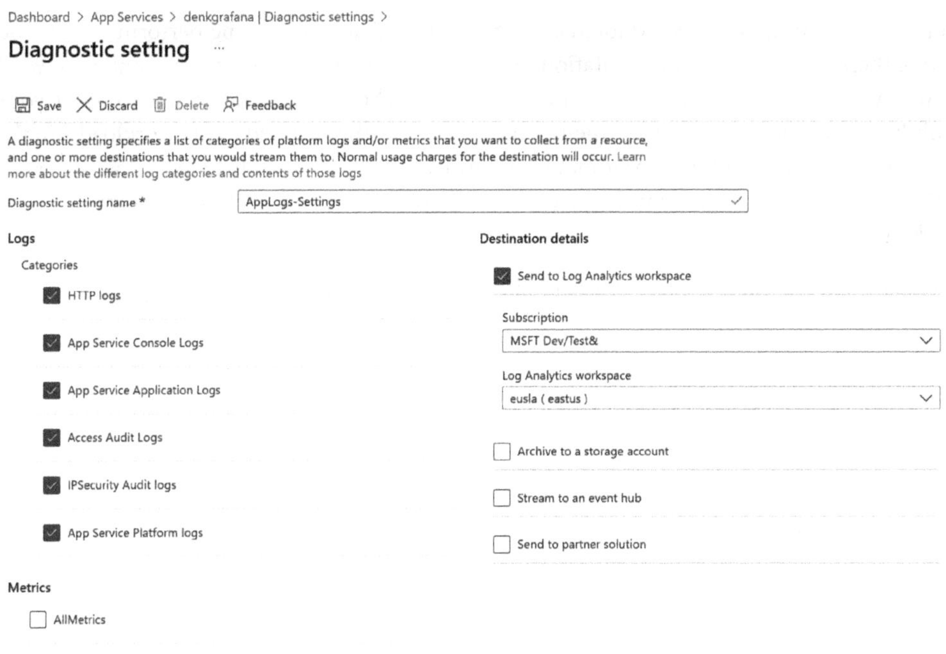

Figure 5.9 – Diagnostic setting

As shown in the preceding screenshot, we can select the logs we need to ingest to the destination. The number of logs impacts the cost of storage. In the case of VMs, if you want to get the logs from the guest OS, you'll need to install either **Microsoft Monitoring Agent** (**MMA**) or **Azure Monitoring Agent** (**AMA**). At the time of writing this book, MMA is set to retire by August 31, 2024. Microsoft recommends upgrading your MMA agent to an AMA agent for existing deployments, and all new deployments should use AMA. The agent will be configured to send the logs to a Log Analytics workspace.

With that, we have completed monitoring. As we have seen in the case of cost efficiency, there will be various trade-offs when it comes to performance efficiency. You can review the trade-offs here: https://learn.microsoft.com/en-us/azure/architecture/framework/scalability/tradeoffs.

# Summary

In this chapter, we covered the third pillar of the WAF, that is, performance efficiency. We started the chapter by looking at the core principles of performance efficiency: design for scaling, performance testing, and performance monitoring. We then reviewed checklists of the core principles. The checklist for design included key areas such as application design, data management, and implementation principles. After that, we discussed performance testing and covered load testing, stress testing, and multi-region testing. Also, we explored different tools available for testing performance. If you incorporate these design and implementation principles into your Azure environment, you won't need to worry about performance degradation ever again. The recommendation is to order your workloads based on priority and verify whether these principles are applied to the workloads. In the last section, we covered monitoring. Without monitoring, we will not be able to detect performance issues. We need to collect logs from the platform, infrastructure, and application to understand bottlenecks and downtime.

In the next chapter, we will learn how we can build reliable applications by leveraging the *reliability* pillar of the WAF.

# 6

# Building Reliable Applications

In the previous chapter, we covered performance efficiency where we saw how we can make sure that our application withstands sudden demand. Now, we will start covering another pillar of the WAF – that is, **reliability**. Organizations offer **service-level agreements (SLAs)** and **service-level objectives (SLOs)** to their end users. Applications should meet these commitments, and with reliability, we are ensuring that these commitments are met. By incorporating reliability into our architecture, we make sure that our workloads are available and can recover from catastrophic failures regardless of the scale of the failure. Compared to the previous chapter, this chapter will be short as we have covered most of the testing and monitoring solutions before.

## Introducing the reliability pillar

In general, the reliability pillar focuses on the following objectives:

- Ensuring you have a highly available architecture and avoid any **single point of failure (SPOF)**
- Business continuity and disaster planning in case of data loss, downtime, or catastrophic failures
- Testing the high availability and recovery of workloads

Building reliability in the cloud requires a shift in mindset, as we learned at the beginning of this book. For example, in traditional application development, the prime focus was on increasing the average time between system breakdowns. This metric is called the **mean time between failures (MTBF)**. The work was primarily devoted to attempts to stop the system from failing. In the cloud, we have distributed systems and the approach will be slightly different, and we will require a shift in our mindset because of the following factors:

- The complexity of distributed systems is high and a single failure in these systems can lead to a domino effect.
- In the cloud, providers use commodity hardware to offer services for a cheaper rate to customers. For this type of hardware, anticipate occasional failures.

- Applications may become temporarily unavailable or throttle if you have a dependency on external services.

- In the modern world, users expect an application to be available 24/7 without any downtime.

We need to accommodate these factors when designing workloads for reliability. In other words, all applications should be designed to anticipate failures and strategies to recover from these failures. To address this, Microsoft Azure has a lot of built-in reliability measures such as the following:

- With the help of **zone-replication** and **geo-replication**, we can replicate Azure Storage, Azure SQL Database, and Azure Cosmos DB out of the box across availability zones and regions.

- Different **storage racks** are used for Azure Managed Disks to protect your disks from hardware failures.

- **Availability sets** are offered to Azure VMs, where we can deploy multiple instances of VMs to different **fault domains** in a data center. Fault domains represent a set of hardware in a data center that shares a common power source, network, and cooling system. Since the VMs are deployed across multiple fault domains, they are protected from hardware failures. Besides fault domains, we also have update domains in availability sets, which will protect the VMs from update and maintenance tasks in the data center.

- An availability set is defined at the data center level; if the entire data center goes offline, availability sets cannot protect your workloads. With the help of **availability zones**, we can deploy VMs across multiple zones. Each zone represents a physically separate location within a region with an independent power source, network, and cooling system. The zones are connected to each other using low-latency optical fiber, enabling faster communication between the zones. If one zone is affected, we will have our VM running in another zone without any downtime. Unless and until the entire region goes offline, your VMs are resilient.

- For **regional redundancy**, we have Azure Site Recovery. We can use this to replicate our VMs in another region and bring them up in case of a disaster.

Now that we know reliability is built into the platform, we shouldn't stop building **resiliency** into our application. Based on your architecture, you need to implement reliability at all levels. We need to plan tactical and strategic mitigations. For example, in case of transient network failure, we need to retry the connection; this is a **tactical mitigation**. On the other hand, failing over your application to another region requires a **disaster recovery** (**DR**) strategy and we can categorize this as a **strategic mitigation**. If we consider the frequency of these mitigations, tactical mitigation will be very common and frequent while strategic mitigation will be planned and unusual, as we don't face regional failures every day, for example. At the end of the day, it all boils down to finding the issue and mitigating it. In order to find these issues, you need to have monitoring and diagnostics set up in your environment. With the help of logs and metrics, we can analyze the issue and find the root cause.

On that note, let's delve deeper into this topic; we will follow the same pattern that we followed in the previous chapter. We will discuss the various reliability principles, design aspects, testing, and reliability monitoring. Let's begin with the principles that are considered critical for assessing the reliability of an application hosted on Azure.

# Reliability principles

As mentioned earlier, if you are planning to build reliability, then you need to take a different route than you follow in traditional hosting. In on-premises, usually, we purchase multiple redundant types of hardware to decrease the impact of hardware failure and increase redundancy. Since this hardware is high-end and expensive, the capital investment is higher. The first step in building reliability in the cloud is accepting the fact that there will be failures and admitting that this is inevitable. All we can do is tailor our plans to mitigate the failures and minimize the impact.

The following principles should be used to assess the reliability of your applications:

- **Understanding the business requirements**: All applications will have specific business requirements, and when we design reliability, the methods that you adopt should reflect these business requirements. For example, say your application is hosted on an Azure VM and the committed SLA to end users is 99.95%; in such a scenario, you can't expect a single VM to cover this. As per the Microsoft SLA documentation, a single VM with a Premium Disk can offer only 99.9%, which is less than the business requirement. In this scenario, we need to improvise our architecture to use multiple VMs in availability sets that can offer 99.95%, matching the business requirement. As there is no cost difference between availability sets and availability zones, you can also deploy multiple VMs across zones and increase the SLA to 99.99%. Though the VMs' cost is the same in both setups, that might not be the case in the case of other services, so check the cost implications and find the right configuration meeting your business requirements.

- **Anticipating failures**: Failures are something that we cannot avoid in cloud environments, and we need to acknowledge this fact. Having said that, while designing for reliability, we need to consider the reliability options for each component in our architecture and ensure that none of the components are a SPOF.

- **Monitoring health**: In order to mitigate issues before they turn into bigger issues, we need monitoring. Every component in our architecture should be monitored, thus ensuring all components are healthy and ready to serve requests. Azure Monitor offers a plethora of tools for monitoring the health of your application and infrastructure.

- **Promoting automation**: As we have seen in the case of the *Operational Excellence* pillar, if we don't automate solutions, we will have to deal with human errors and inconsistencies in configuration. For example, if you are using Azure Site Recovery to fail over your VM to a secondary region, by default, that's a manual process. If we automate this using automation runbooks, we can failover automatically based on metrics.

- **Self-healing**: As the name implies, self-healing refers to the capability of a system to deal with failures and remediate them automatically. This is an advanced topic and requires high-level automation and monitoring to achieve self-healing. The aim of self-healing is to maximize reliability from the beginning.

- **Scaling out**: We have already seen the advantages of scaling out for the *Performance Efficiency* and *Cost Optimization* pillars. When we design for scale-out, based on the demand, the number of instances is increased or decreased. In addition to improved performance and optimized cost, scaling out can enhance the overall reliability by handling expected and unexpected loads.

Now that you are familiar with the reliability principles, let's move on to the design aspects of reliability. When you design solutions aligned with the reliability pillar, you should incorporate these design aspects into your architecture.

## Designing for reliability

Microsoft has developed a **design checklist** to ensure the reliability of workloads in Azure. Applications should maintain a percentage of uptime based on the business requirements; this is what we call **availability**. We need to find the right balance between high resiliency, low latency, and cost. If we focus on a single factor and improve it, then it can have repercussions on other factors. For example, you can deploy multiple VMs to improve resiliency, but this will increase the cost. In a nutshell, we need to find a balance between these factors. The following is the checklist provided by Microsoft for designing reliable applications:

- Define availability and recovery targets to fulfill your business needs

- Capture needs to incorporate resiliency and availability into your applications

- Ensure that the application and data platforms meet your reliability requirements

- Set up connection paths to enhance the availability

- Set up availability zones to improve reliability

- Validate SPOFs in the architecture

- Define the outcomes if an SLA breach happens

- Validate whether applications can operate even if there is a dependency failure

In the upcoming sections, we will cover these points in depth. The following Azure services can be used in your architecture to enhance reliability:

- Azure Load Balancer

- Azure Application Gateway

- Azure Front Door

- Azure Traffic Manager

- Azure Kubernetes Service

- Azure Site Recovery

- Azure NAT Gateway

- Azure Service Fabric

As you can see, most of the services are load balancers, which is inevitable in building any reliable applications. Beyond this, Azure Kubernetes Service, Azure Service Fabric, Azure Virtual Network NAT, and Azure Site Recovery can be used for application resiliency.

Now will explore the checklist components one by one, starting with target and non-functional requirements.

## Design requirements

Availability targets and recovery targets are metrics that allow you to assess the uptime and downtime of your application. You need to define these metrics and ensure that they match your business requirements. Having a clear definition of these metrics will help us set a goal and assess our current setup against the defined goal. Apart from these targets, there are many other considerations we should consider for improving the overall reliability and availability of the application, which we will cover in the upcoming sections.

The following key points should be considered while defining target and non-functional targets:

- Define the acceptable level of downtime for your application

- Define the maximum data loss that can be accommodated in case of a disaster

- Define the application and data requirements to enhance the reliability of your workload

- Define monitoring to capture the overall health of your application

We need to focus on two things; one is recovering from failures and the other is maintaining the healthy state of your application without major downtime. Based on the key points discussed in this section, we will cover three main concepts: an **availability target**, a **recovery target**, and **platform requirements**. Let's start with availability targets.

### Availability targets

You might have heard of the SLA of an application. This is an availability target that defines the commitment that application providers agree with the end user for the performance and availability of the application. This will be defined as an uptime percentage; for example, if you see a 98% SLA for an application, that means the application provider guarantees 98% of uptime in each time period. If the provider fails to meet the commitment, then that will be a breach of the agreement, often referred to as an **SLA breach**. In the agreement itself, there will be norms defined on what needs to be done in case of an SLA breach. When you are designing an application, you should ensure that availability and reliability are built-in to the architecture.

In order to make sure that the availability metrics are attained, we need to monitor the application. The key targets are as follows:

- **Mean Time Between Failures** (**MTBF**): The average time between failures of a component in your application
- **Mean Time to Recover** (**MTTR**): The average time taken to restore a component after failure

We need to define these metrics for the application and it should be monitored to ensure these values are attained. Now that you know about availability targets, let's talk about recovery targets.

### Recovery targets

While availability targets define the commitments around the performance and availability of your workload, a recovery target defines how long the application can be unavailable and how much data loss is acceptable in the event of a disaster. As we have seen in the case of availability targets, here we have two metrics that need to be defined, which are as follows:

- **Recovery Time Objective** (**RTO**): The maximum acceptable downtime for an application after an incident
- **Recovery Point Objective** (**RPO**): The maximum acceptable duration of data loss during a disaster

Recovery targets are non-functional requirements of the workload and should be verbalized by business needs. The values for the preceding metrics should be defined by the organization based on their business requirements. Beyond that, we need to plan our business continuity and DR strategy to align with these metrics. Ideally, you should conduct DR drills periodically to verify that the DR plan is working and to confirm that desired RTO and RPO values are maintained.

With that, we have completed the recovery targets, so now we will move on to the platform requirements.

### Platform requirements

Microsoft Azure offers built-in resiliency and high-availability features to support your application hosted on it. Having said that, based on the SKU that you select, the reliability features and SLA may vary. It's recommended that you review the service documentation to understand the different tiers of service and their corresponding features and SLAs before deployment. By doing so, you can be sure that the service tier meets your application requirements. For example, you are planning to use Azure Storage to store your application data as blobs. The application requires storage durability of 99.9999999999% (12 nines) over a given year. If you are choosing **Locally Redundant Storage** (**LRS**), then the durability offered by Azure is 99.999999999% (11 nines) durability of objects over a given year. The durability value doesn't match the application requirements, so we need to explore the next tier. Luckily, **Zone Redundant Storage (ZRS)** offers durability of 99.9999999999% (12 nines) over a given year, which is exactly what our application needs. Similarly, we need to capture the workload requirements and validate whether the requirement matches the features and SLA offered by the Azure service. In this way, we can make sure that the application is hosted on a platform that matches the requirements. Besides the SLA and features, Azure also provides platform features to configure high availability and business continuity, such as the following:

- **Paired regions**: Technically, if you are setting up availability zones, it will be cheaper and less complex. At the end of the day, if the application requires global redundancy, then you must deploy it across multiple regions. When choosing the regions, you should focus on the concept of paired regions in Azure. They offer a lot of benefits when it comes to DR, such as priority in the recovery sequence, sequential updating, physical isolation, data residency, and so on. However, since the paired regions are part of the same geography, you need to worry about the data residency and compliance requirements. Furthermore, Microsoft pushes platform updates to the paired regions in a sequential manner so that one of the regions is always available during the planned maintenance. To learn about the paired regions, you can review the following resource: `https://learn.microsoft.com/en-us/azure/best-practices-availability-paired-regions`.

- **Availability zones**: When deploying resources, we can deploy them to a single zone or across multiple zones to improve high availability. As we discussed earlier in this chapter, availability zones are physically separated data centers connected using low-latency optical fiber. If you deploy your application across multiple zones, even if a single zone fails, your application can run in another zone.

- **Availability sets**: Using an availability set, we can deploy our VMs across fault domains and update domains. Availability sets can protect VMs from hardware failures within a data center.

We need to include these platform design aspects when building solutions for reliability. Now, we will move on to data platform requirements.

## Data platform requirements

We need to make sure that the data platform is highly available with the help of features available in Azure. As discussed earlier, reliability features might not be available in certain SKUs/tiers. For example, Azure SQL Database General Purpose and Azure Storage LRS offer three replicas within the same data center. On the other hand, Azure SQL Business Critical and Azure Storage ZRS come with three replicas across availability zones. We need to focus on the following points when it comes to the data platform requirements:

- **Data consistency**: Decide what level of consistency is required for your data. Based on the consistency requirements, data types should be categorized. Achieving consistency essentially involves a trade-off between availability and partition tolerance. To achieve this, the data will be distributed across multiple zones or regions and the data will be pulled from the closest source to shrink the latency and response time. For example, Azure Cosmos DB offers five consistency levels: *Strong*, *Bounded staleness*, *Session*, *Consistent prefix*, and *Eventual*. You need to establish which consistency is right for your application based on your requirements.

- **Replication and redundancy**: In case of failure, our database solution should be available to serve requests from the application. Furthermore, we should be able to recover the database if one of the available restoration points is corrupted. For example, in Azure SQL Database and Azure Cosmos DB, we have the option to replicate to another region. For Azure Cosmos DB, we have an automatic failover feature to fail over to another region if the primary region fails. Similarly, we have SQL Database Active's geo-replication feature for replicating your data to a secondary region.

These factors should be considered when designing reliable databases in Azure. Now, let's take a look at the networking and connectivity requirements.

### Networking and connectivity requirements

Networking and connectivity are very important in application design. Even if you have redundancy set up for the infrastructure, without designing reliable networking and connectivity, we can't deliver resiliency. The following aspects should be considered when designing networking and connectivity solutions:

- **Using global load balancing solution**: In Azure, Azure Front Door, Azure Traffic Manager, or Azure CDN can be used to route requests to applications that are deployed across multiple regions. Keep in mind that Azure Traffic Manager is a DNS-based load-balancing solution, which means you need to wait for the DNS propagation to occur. We need to set up low TTL values for our DNS records to tackle this; however, not all service providers may honor this. You can take a look at the reference architecture provided by Microsoft at `https://learn.microsoft.com/en-us/azure/architecture/example-scenario/hybrid/hybrid-cross-cloud-scaling`.

    If you prefer transparent failover, you can go with Azure Front Door. The reference architecture for Front Door is available at `https://learn.microsoft.com/en-us/azure/frontdoor/front-door-overview`.

- **Setting up redundant connections from on-premises**: We can set up connectivity to on-premises using Azure ExpressRoute or VPN. While setting up connections, we need to set up redundant connections to avoid uninterrupted connectivity. The recommendation is to configure active-active configuration to facilitate connection if one connection goes offline. You can learn more about the active-active setup for Azure VPN Gateway at `https://learn.microsoft.com/en-us/azure/vpn-gateway/vpn-gateway-highlyavailable#dual-`

`redundancy-active-active-vpn-gateways-for-both-azure-and-on-premises-networks`.

- **Setting up failure paths for hybrid connectivity**: Even if you have a redundant ExpressRoute connection, we need to set up a Site-to-Site VPN as a failover path. In case **ExpressRoute connectivity** fails, we can rely on the Site-to-Site connection for **hybrid connectivity**. The reference architecture is available at `https://learn.microsoft.com/en-us/azure/architecture/reference-architectures/hybrid-networking/expressroute-vpn-failover`.

- **Eliminating SPOF**: In your architecture, eliminate all SPOFs. This is the first step in building reliability. For example, if you are using a **Network Virtual Appliance** (**NVA**) for inspecting packets, instead of a single NVA, use multiple instances behind a load balancer. You can see the reference architecture at `https://learn.microsoft.com/en-us/azure/architecture/reference-architectures/dmz/nva-ha#load-balancer-design`.

Furthermore, we need to leverage the zone redundancy features available in network and connectivity solutions. The following factors should be considered for zone-aware services:

- Azure offers the capability to deploy zone-redundant virtual network gateways. By doing so, the gateway instances will be deployed across availability zones, protecting your instances from data center failures. You can learn more about zone-redundant gateways at `https://learn.microsoft.com/en-us/azure/vpn-gateway/about-zone-redundant-vnet-gateways`.

- **Using zone redundancy in Application Gateway**: If you are using Application Gateway v2, you can opt for zone-redundant configuration and autoscaling. With Standard v2 instances, Application Gateway will be deployed across availability zones; this will improve the reliability of the service. More details can be found at `https://learn.microsoft.com/en-us/azure/application-gateway/overview-v2`.

- **Using Load Balancer to route traffic to availability zones**: When we are deploying VMs across multiple availability zones, we need a solution to route the request between the VMs. This is where we need to use Azure Load Balancer, as shown in the following figure; here, we can see that the requests from the end users are routed to the backend VMs deployed across availability zones with the help of Load Balancer:

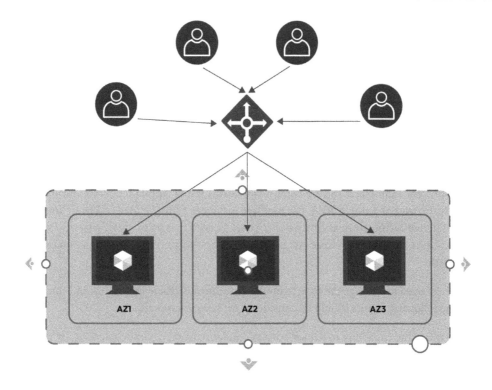

Figure 6.1 – Load-balancing to instances

- **Configuring health probes for Load Balancer**: In the preceding figure, we have Load Balancer, which routes the request to backend instances. If we don't configure health probes, Load Balancer will route the requests to the backend server even if they are unhealthy. With health probes, Load Balancer will periodically check the health of instances, and after a consecutive number of failures, Load Balancer will stop routing requests to unhealthy servers.

- **Monitoring critical application endpoints**: We need to set up custom health probes for our application endpoints and dependent services. This will help us constantly assess the health of the application and stop routing requests if the endpoint is not healthy.

With that, we have completed the requirements for reliability, and now we will look at the key points of application design.

## Application design

As mentioned earlier in this chapter, building reliable applications in the cloud is completely different from the traditional way of application deployment. The first and foremost thing we need to acknowledge is that failures can happen in the cloud. So, rather than trying to stop something unavoidable, we should focus on designing solutions that can overcome these failures. In this section, we will quickly review the key points in application design. Let's cover the key points:

- **Using availability zones**: You are already aware of availability zones and how these zones can improve the reliability of your application by deploying its instances across availability zones. You should consider multi-region deployment only if availability zones cannot fulfill your application target metrics; in all other scenarios, the architecture should be designed as zone-redundant. Deploying applications across multiple regions will have cost implications, such as it requiring additional bandwidth, networking, and so on.

- **Responding to failures**: As we discussed, avoiding failures in the cloud is not possible. All we can do is transform our application to handle these failures. When we design an application, we need to design it in such a way that it can tolerate hardware, zonal, regional, service-level, and dependency failures. An active-active configuration will seem expensive, but it is the right thing to do if you want to achieve reliability.

- **Using managed services**: When choosing solutions, always see whether a PaaS solution can fulfill your requirement. The reason is most PaaS solutions will have built-in high availability and we can leverage this to build reliability. Unlike IaaS solutions, we don't have to worry about managing or maintaining infrastructure, as it will be Microsoft's responsibility in the case of PaaS solutions.

- **Designing for scale-out**: As we have seen in the case of performance efficiency, we need to design for scale-out. Based on the number of requests, we will be able to increase or decrease the number of instances, which, in turn, can have a positive effect on the cost, performance, and reliability of the applications.

You can read about these topics in depth at `https://learn.microsoft.com/en-us/azure/architecture/framework/resiliency/app-design`.

Now that we have discussed application design, let's talk about resiliency and dependencies.

## Resiliency and dependencies

As mentioned earlier in this chapter, you should treat reliability design as part of your architecture design so that you can design reliable solutions. This mindset will help you avoid SPOFs and eliminate risk. Applications will have dependencies, and if a dependency is not functioning, that will affect the overall functioning of an application. Assume that you have a three-tier application; the frontend and mid tier are working, and the data tier is malfunctioning. In this scenario, the requests from end users will be accepted by the frontend and then routed to the mid tier; however, the data tier will not be able to commit the transactions to the database, as it is not working. Basically, users cannot complete the actions due to the absence of the data tier and your SLA will be breached if the data tier is not fixed immediately. The bottom line is that the reliability of all application components, including their dependencies, is important.

When it comes to dependencies, we have both internal and external dependencies. Components that are required for the proper functioning of the application and are within the application scope are termed **internal dependencies**. On the other hand, as the name implies, **external dependencies** are outside the scope of an application, such as third-party services or another dependent application. When designing, you should make a list of all dependencies. Common dependencies are Azure Active Directory, ExpressRoute, a VPN, or a central NVA, as well as APIs that your application interacts with.

Microsoft has highlighted some key points when it comes to resiliency and dependencies; they are as follows:

- **Using Failure Mode Analysis (FMA) to build resiliency**: FMA is a stepwise practice designed to capture fault symptoms that happen before or after a system failure. In this procedure, the failures are categorized and prioritized based on their severity and frequency. You should incorporate an FMA process as part of your architecture and design stages. By doing so, you ensure that failure recovery is unified from the beginning. The idea is to eliminate all SPOFs, as these points can lead to the breakdown of the entire application. SPOFs are a flight risk for your application, as they are heavily prone to failures and will cause outages.

- **Scaling the impact of an outage with each dependency**: Strong dependencies play a decisive role in an application's functionality and readiness. If the dependency is unavailable, that will affect the availability of the application. Meanwhile, weak dependencies will not affect the overall health of the application. Even with the failure of the dependency, the application can still function. We need to rank the dependencies based on their strengths. If the dependency is strong, we need to ensure that the dependency is reliable.

- **Maintaining SLAs for dependencies**: Understand the SLA of dependencies and ensure that they align with the SLA of the application and attain the availability targets of your application. By knowing the SLA, we can add additional reliability to the dependency if the SLA is lower than the target.

- **Confirming that applications can operate in the absence of their dependencies**: Having a strong dependency means that your application cannot operate if the dependency fails, which, in turn, affects your availability targets. You should make efforts to minimize dependencies to achieve full control over your application's reliability.

- **Decoupling an application's life cycle from dependencies**: The operational agility of your application will be affected if the life cycle of the application is tied to the life cycle of the dependencies. The best practice is to decouple the application life cycle from dependencies. Nevertheless, this recommendation applies when you are making new releases.

Microsoft has developed a set of best practices for designing reliable applications, which are available at `https://learn.microsoft.com/en-us/azure/architecture/framework/resiliency/design-best-practices`. All points mentioned in the documentation are excerpts from the design recommendations we have covered so far. With that, we have concluded the design section and will move on to testing.

# Reliability testing

Periodic testing should be conducted to revisit your design decisions. During testing, we need to confirm the availability and recovery targets are met. If there is a need to improve your architecture, then you should do that. Key Azure services that will be used for building reliability are as follows:

- **Azure Site Recovery**: To replicate VMs to a secondary region and in case of a disaster, we can easily bring up the VM from the DR region.

- **Azure Pipelines**: In the *Operational Excellence* and *Performance Efficiency* chapters, we saw how Azure Pipelines can be used to implement continuous testing, continuous integration, and continuous delivery. With the help of pipelines, we can validate the reliability of the code and delivery.

- **Azure Traffic Manager**: This is a DNS-based load balancer that offers different routing methods, such as priority, weighted, performance, and geography routing, which can be used with applications deployed across multiple regions.

- **Azure Load Balancer**: At the time of writing this book, Azure Load Balancer is a Layer-4 load balancer used for load balancing requests to VMs deployed in a virtual network. Having said that, we also have a global load balancer in preview, which can be used for applications deployed in multiple regions.

- **Azure Front Door**: This is a global solution that can be used for transparent failover for applications deployed across multiple regions.

- **Azure Application Gateway**: This is a Layer-7 load balancer that supports multiple backend services such as Azure VMs, Azure App Services, App Service Deployment Slots, or even external services deployed outside Azure.

On these services, we need to test failover to ensure that they are working as expected and that the availability metrics (MTBF and MTTR) and recovery metrics (RPO and RTO) are met. Some reference architectures that you can refer to include the following:

- *Failure Mode Analysis for Azure applications* (`https://learn.microsoft.com/en-us/azure/architecture/resiliency/failure-mode-analysis`)

- *High availability and disaster recovery scenarios for IaaS apps* (`https://learn.microsoft.com/en-us/azure/architecture/example-scenario/infrastructure/iaas-high-availability-disaster-recovery`)

- *Back up files and applications on Azure Stack Hub* (`https://learn.microsoft.com/en-us/azure/architecture/hybrid/azure-stack-backup`)

The following checklist can be used to verify whether your applications are designed with reliability:

- Test the environment periodically to substantiate the existing configuration, thresholds, targets, and assumptions

- Automation can be used to remove human errors and configuration drift and perform testing with as much automation as possible

- Maintain the development environment as a reflection of the production environment and perform testing on both environments

- Inject faults and verify how the application handles them

- Develop DR plans and validate whether the DR failover working as planned

- Design applications to run with minimal functionality in the DR region

- Design a backup plan to back up your critical application data to handle catastrophic failures

- Validate your ExpressRoute and VPN connections and ensure that redundant connections and failback are working

- Configure timeout between inter-component calls

- Develop retry logic to handle transient application failures or dependency failures

- Implement health probes on your load-balancing solutions to check the health of your backend servers or endpoints

- Apply chaos engineering principles

- Monitor failed backup jobs and retry backups

That's all we have for testing; now we will move on to the last topic in this chapter, monitoring.

## Monitoring

By now, you know that monitoring is something inevitable regardless of which WAF pillar we are dealing with. Basically, we need insights into component failures, why the component failed, and when it failed. We have already seen how Azure Advisor and Azure Monitor play a crucial role in capturing the recommendations and insights related to the pillars of the WAF.

The following key points should be considered for monitoring the reliability of applications:

- Use Application Insights to validate the availability of applications to ensure the availability metrics are met.

- Collect application logs using Azure Monitor so that we can troubleshoot any application-related failures.

- Ensure all application components are monitored and data ingested to a centralized location such as Azure Log Analytics.

- All key metrics should be captured and plotted on dashboards for the applications and infrastructure teams to monitor. You can use Azure Workbooks or Grafana to build dashboards.

- The health model should be defined using thresholds and targets; the values should be validated against the availability and recovery targets.

- Monitor Azure Service Health events to be aware of platform-related issues. Alerts should be set up so that your team is notified about planned maintenance and outages on the Azure platform.

- Configure Azure Resource Health to review the health of resources in your Azure subscription. Administrators should be notified in case of failure events and health state changes.

- Long-running workflows are prone to failures, so these workflows should be monitored for failures and alerts should be set up.

We have already covered the key services that can be used for monitoring applications in previous chapters. Here, we will list the services and add snapshots again for your reference:

- Azure Monitor:

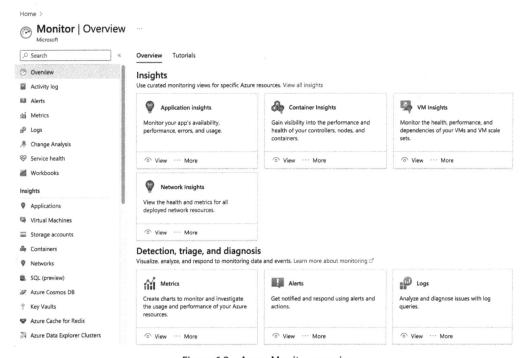

Figure 6.2 – Azure Monitor overview

- Azure Application Insights:

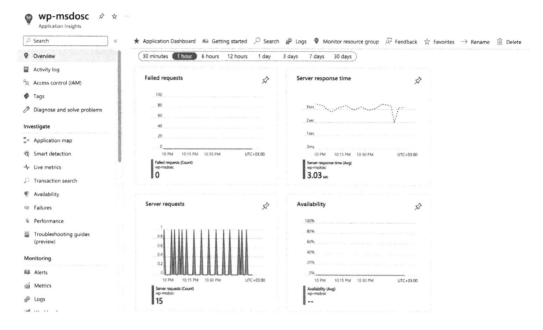

Figure 6.3 – Application Insights

- Azure Service Health:

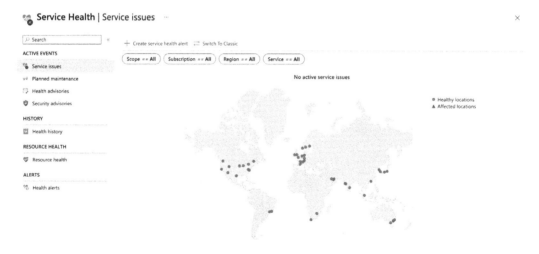

Figure 6.4 – Azure Service Health

- Azure Resource Health:

Figure 6.5 – Azure Resource Health

- Azure Advisor:

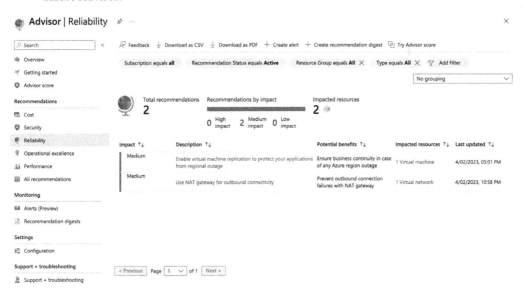

Figure 6.6 – Reliability recommendations in Azure Advisor

There are couple of reference architectures that you can refer to configure monitoring for your application:

- Hybrid availability and performance monitoring (https://learn.microsoft.com/en-us/azure/architecture/hybrid/hybrid-perf-monitoring) using Azure Monitor

- Unified logging for microservices applications (`https://learn.microsoft.com/en-us/azure/architecture/example-scenario/logging/unified-logging`)

With that, we conclude this chapter. We hope that it gives you insights to build reliable applications in Microsoft Azure.

## Summary

In this chapter, we focused on the reliability pillar of the WAF. After introducing this concept, we discussed the principles of reliability. When you are developing reliable solutions in the cloud, you need to acknowledge the fact that in the cloud, failures will happen. Without wasting our time trying to prevent something we cannot avoid, we need to focus on building solutions that can handle these failures and remain reliable. We covered the key design aspects, such as the platform requirements, application design, and dependency management. Furthermore, we discussed how we can improve the reliability of the application and monitor failures.

In the next chapter, we will see how we can build secure solutions in Azure using the principles outlined in the *security* pillar of the WAF.

# 7

# Leveraging the Security Pillar

In on-premises systems, the data center and infrastructure are fully managed by the organization, and it is the organization's responsibility to secure its environment. **Security** is a complex topic, and it raises a lot of questions when the organization adopts a public cloud such as Microsoft Azure. Nowadays, attacks are getting more sophisticated, and attackers are improvising their techniques. So, we need to increase the security landscape of our environment to cope with these attackers. Attackers target vulnerabilities in system configurations, operational methods, and the way the users handle their devices. People who want to exploit your vulnerabilities will constantly scan your environment for security gaps; at the right moment, they will penetrate and compromise your environment. These days, most organizations store customer data, and when this data is compromised, it ruins your reputation. With the help of best practices outlined in the security pillar of Microsoft's **Well-Architected Framework (WAF)**, we will discuss how to design solutions with security and protect our environment from attackers.

Security is the key aspect of every architecture regardless of whether your application is running in the cloud or on-premises. You might have heard of the **CIA triad**, which stands for **confidentiality**, **integrity**, and **availability**. This triad is used as a model for the development of secure solutions. With the help of the model, we can detect vulnerabilities and mitigate them. While designing solutions, we need to ensure that all three pillars of the CIA triad are fulfilled, thus ensuring your organization has a strong security landscape and can handle security events effectively. If you don't take the CIA triad seriously, your solutions will have security gaps and can adversely affect your business operations, revenue, and organization's status. In this chapter, we will discuss the key architectural design considerations and principles regarding the security of solutions deployed in Microsoft Azure.

# Designing secure solutions

When you are designing solutions aligned to the security pillar, make sure you cover the areas shown in the following figure:

Figure 7.1 – Key design areas

In *Chapter 6*, *Building Reliable Applications*, we learned that the first thing we need to acknowledge is that failures can happen in the cloud, and all we can do is find ways to mitigate the failure. Similarly, while dealing with security, always assume a breach or compromise. We always assume that there is a breach of security, and we define controls to mitigate these breaches. Microsoft has developed a framework called the **Zero Trust model** (`https://learn.microsoft.com/en-us/security/zero-trust`), which states, "*Never trust, always verify*." The three fundamental ideas of the Zero Trust model are as follows:

- Always verify explicitly
- Use the principle of least privilege
- Always assume a breach

While designing solutions, we need to ensure that we adhere to these fundamental ideas of the Zero Trust model. When it comes to the cloud, there is a shift in ownership of data centers. In the public cloud, the data center will be owned by the cloud provider (such as Microsoft), and the responsibilities are shared between the customer and the cloud provider. We can streamline the tedious task of securing cloud environments through specialization and a shared responsibility model. Let's understand what these concepts are:

- **Specialization**: In on-premises, it's the organization's responsibility to provide the physical security for the data center, patching of servers in the data center, and hypervisor configuration. By moving to the cloud, organizations hand over these responsibilities to the cloud provider. In this way, organizations don't need to invest in security personnel and infrastructure management teams. Cloud providers will make sure that the necessary security and infrastructure management measures are applied to their data centers in order to stay compliant with industry standards.

- **Shared responsibility model**: As workloads get transferred from customer-managed data centers on-premises to the cloud, there will be a shift in the responsibilities as well. As discussed in the previous point, physical data center security, server patching, and hypervisor updates will be handled by the cloud provider. Apart from these responsibilities, based on the deployment model (IaaS, PaaS, or SaaS), there will be variations on what is managed by the cloud provider and what is the customer's responsibility. The following figure shows the shared responsibility model for each deployment model:

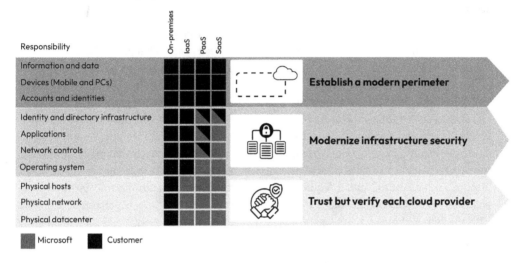

Figure 7.2 – Shared responsibility model

As you can see in the preceding figure, in on-premises, everything is managed by the customer. Once they start using the IaaS solution, the physical hardware becomes Microsoft's responsibility and the rest of the stack is the customer's responsibility. As we shift to PaaS solutions, more responsibility will be on Microsoft's side and customer responsibility is minimized. This is the reason why Microsoft recommends using managed services (PaaS) wherever possible, so the customer can be more productive rather than focusing on infrastructure management. When developing secure solutions, you need to understand the boundaries of your responsibilities and make sure you implement security controls for everything that comes under your responsibility.

Now, let's explore the key areas of the WAF security pillar. These key areas are used to develop the principles and design of the security pillar.

## Key areas and security resources

When you are developing solutions, your focus should be on certain key areas, as defined by Microsoft for the security pillar:

- **Identity management**: Use Azure AD as the identity management solution for authenticating and authorizing the users. Since Azure AD is a fully platform-managed solution, you don't need to manage infrastructure and your developers can easily integrate Azure AD with applications. You can integrate your on-premises AD using existing domains on your own or you can create new domains. Azure AD is not only used by Azure and resources in Azure but also by other SaaS solutions such as Microsoft 365 and Dynamics 365. When it comes to consumer-facing applications, you can use Azure AD B2C, which will help your users to authenticate with social accounts such as Google, Facebook, and LinkedIn.

- **Manage access to infrastructure**: Azure subscriptions will be mapped to an Azure tenant (Azure AD) when they are created. Nevertheless, you can always change the directory of the subscription from one Azure AD to another. Due to the mapping, there is a trust relationship between Azure resources and Azure AD. With the help of Azure **role-based access control** (**RBAC**), it provides access for users to manage Azure resources. While granting permissions, you should always follow the principle of *least privilege*. In other words, you should only give access to users at the required scope (such as management group, subscription, resource group, or single resource) and enough permissions to complete their day-to-day work. The users shouldn't be under-privileged or over-privileged while gaining access to resources. Azure offers built-in roles for RBAC and, if required, you can create your own custom role. Furthermore, you need to audit all changes that are happening to your infrastructure, which can be achieved with the help of Azure activity logs with an extended retention period.

- **Application security**: The application development best practices that you follow on-premises should be followed in the cloud as well. For example, encrypt data in transit using TLS, guard applications against CSRF and XSS attacks, and block SQL injection attacks. If your application requires access keys, secrets, or certificates, then consider using Azure Key Vault rather than directly embedding them inside the code. Moreover, you can leverage managed identities to authenticate to Azure AD and access Azure resources seamlessly. For example, if your application is hosted in App Service, you can create a managed identity for App Service and grant access to Azure Key Vault.

- **Data sovereignty and encryption**: Azure has 60+ regions; make sure you choose the right region to deploy your resources based on your organization's data compliance and residency requirements. Make sure your data resides in the correct geopolitical region when you are persisting data to Azure. In the case of disaster, Azure replicates data to a paired region that is also within the same geography as your primary region. Data services such as Azure Storage, Azure SQL Database, Azure Synapse Analytics, and Azure Cosmos DB support encryption at rest.

- **Network security**: There are native network security solutions available in Azure along with third-party solutions from vendors such as FortiGate, Palo Alto, F5, and so on. In addition to these appliances, we can use Azure Web Application Firewall to protect our web applications from attacks. Moreover, we have Azure Network Security Group, which offers additional security at the subnet and interface level.

Besides the key areas, Microsoft has a set of comprehensive security advice that consists of the following points:

- Deploy **Microsoft Defender for Cloud** (`https://azure.microsoft.com/services/security-center`), which provides security monitoring and policy management for workloads deployed in Azure and non-Azure environments

- Refer to *Security in the Microsoft Cloud Adoption Framework for Azure* (`https://learn.microsoft.com/en-us/azure/cloud-adoption-framework/secure`) for a high-level overview of the cloud security end state

- Leverage the security architecture design (`https://learn.microsoft.com/en-us/azure/architecture/guide/security/security-start-here`) to get reference implementation architectures

- Follow the Azure Security Benchmark (`https://learn.microsoft.com/en-us/security/benchmark/azure`), which comprises best practices and guidelines for Azure security

- Refer to end-to-end security documentation (`https://learn.microsoft.com/en-us/azure/security/fundamentals/end-to-end`), which will help you get familiar with security services in Azure

- Apply the top 10 security best practices for Azure (`https://learn.microsoft.com/en-us/azure/cloud-adoption-framework/secure/security-top-10`) to improve your security landscape

- Embrace Microsoft cybersecurity architectures (`https://learn.microsoft.com/en-us/security/cybersecurity-reference-architecture/mcra`), which consist of comprehensive information on Microsoft security abilities and integration with third-party applications

With that, we will discuss the principles of the Azure WAF security pillar.

## Understanding the security pillar principles

Unlike other pillars, the security pillar principles can be applied to on-premises and cloud resources. The only difference is the alignment in the shared responsibility model we discussed in *Figure 7.2*; apart from that, the basic concepts are the same. The CIA triad is the foundation of the principles we are going to cover. Based on the principles, we can draw the following conclusions:

- They provide context for questions related to security. They allow us to determine the relevance or importance of specific aspects in relation to security.

- They also help establish the relationship between various aspects and security.

As described in the case of the other pillars that we've covered in previous chapters, the following principles should be used as lenses to assess the security of your workloads:

- **Select resources and plan how to harden them**: In the cloud, we have multiple services to host your application and its dependencies. Once we choose the right service, we need to plan how to harden the service. Remember the shared responsibility model; if you choose to go with an IaaS solution, that means you must handle the entire hardening process. Let's say you plan to proceed with a PaaS solution, then most of the hardening will be handled by Microsoft; all you need to handle is the data security and encryption.

- **Automate and follow the principle of least privilege**: Always grant permissions using the principle of least privilege. In simpler terms, a user should be granted an adequate number of permissions on a scope (such as subscription, resource group, or individual resource). The permissions should empower the user to accomplish their day-to-day tasks. While assigning the permissions, they shouldn't be under-privileged or over-privileged. If the user is under-privileged, they will not be able to complete their day-to-day duties; on the other hand, if they are over-privileged, they will possess authorization to perform tasks that are out of their daily tasks. So, in short, always give the right set of permissions. Furthermore, integrate automation through DevSecOps to include security controls automatically without human intervention.

- **Data classification and encryption**: Classify the data based on the sensitivity of the data. For example, classify data as general, public, confidential, and so on. Based on the sensitivity, assign the right security measure to avoid data leaks and exposure to unwanted audiences. Besides data classification, apply encryption to data at rest and in transit using industry-standard encryption technologies. All the encryption keys and certificates should be stored securely in solutions such as Azure Key Vault for better auditing and management.

- **Security monitoring and incident management**: Use SIEM solutions such as Microsoft Sentinel to track security events by ingesting data from your solutions. Microsoft Sentinel offers playbooks and automation to establish automated incident management and mitigation. Furthermore, integrating with Azure Monitor offers capabilities to query data and correlate events that are happening across your applications.

- **Endpoint protection**: Deploy firewalls and web application firewalls to protect your external and internal endpoints. External endpoints are more vulnerable to attacks, so they should be protected with different layers to avoid penetration. In Azure, we have Azure Firewall and **Network Virtual Appliance** (**NVA**) for protecting your network integrity. Furthermore, we have **Web Application Firewall** available in services such as Azure Application Gateway, Azure Front Door, and Azure CDN to protect your web applications. In order to protect your virtual networks, Azure also offers **distributed denial of service** (**DDoS**) protection to tackle DDoS attacks.

- **Defend against code-level vulnerabilities**: Cross-site scripting and SQL injection should be identified and mitigated before we ship the code to production. Include security fixes, code base patching, and dependency patching in the operational life cycle itself.

- **Threat modeling**: Microsoft recommends establishing techniques to capture and mitigate known threats. Use techniques such as penetration testing, static code analysis, and code analysis to verify threat mitigation and detect and prevent future vulnerabilities.

With that, we are concluding the security principles. Treat these principles as your golden rules while designing for security. Now, we will move on to design areas, where we will cover how to design for governance, identity and access management, networking, data protection, and applications.

# Design areas

When we design an end-to-end solution with security in mind, we must focus on multiple areas. We cannot just focus on infrastructure security and leave application security. All the areas we are going to cover in this section should be treated based on the return on investment, as a small misconfiguration can lead to security gaps. The following security areas will be covered in this section:

- Governance
- Identity and access management
- Networking
- Data protection
- Application and services

Let's start with the first one on the list: governance.

## Governance

The first design area we consider is **governance**. In this section, we will cover the governance checklist, the Azure Security Benchmark, and the reference architecture shared by Microsoft. Governance is about enforcing compliance and measures to check whether the organization is meeting the organizational requirements while deploying resources to Azure. With the help of Azure Policy, we can enforce industry standards such as ISO27001, PCI-DSS, and NIST. Azure has built-in initiatives that cover most of the industry standards and we can ramp up our workloads with a few clicks. In governance, we focus on the following parameters:

- **Define**: Develop organizational standards for operations, technologies, workloads, and configurations based on your internal and external factors. For example, if you are handling customer credit card information, you need to have PCI-DSS compliance. Similarly, if your company is based in Europe, you will have data residency and sovereignty requirements as per GDPR. These kinds of factors help you define your organizational standards. Once the factors or requirements are identified, they need to be documented so that we can make sure all our deployments are aligned with these standards.

- **Improve**: As business requirements change, standards defined initially might need to be changed. We need to evaluate requirements continually to make sure that our documented requirements align with the current organizational requirements.

- **Maintain**: With the requirements in place, we need to make sure that we monitor and audit them from time to time. Without monitoring, we would not be able to determine that security posture.

When we roll out security practices, we need to invest based on the criticality of the systems. We need to invest more in high business impact and highly exposed systems. For development systems, use synthetic data such as production data. By doing so, we can make sure our testing is done in a production-like environment. Having said that, we don't need to invest in the development environment as much as we invest in the production environment as the development environment is a less critical system and doesn't come with any potential data exposure. Microsoft CISO Workshop (`https://learn.microsoft.com/en-us/security/ciso-workshop/ciso-workshop`) – *Module 4a* – provides a comprehensive prioritized list of security initiatives that can be adopted to improve the security landscape of your organization.

Now, let's take a quick peek at the checklist for governance. The following considerations should be made while designing for governance:

- Implement Azure Landing Zone for your workloads. Landing Zone provides subscription democratization and helps you to isolate environments and apply different security controls to different environments. The Landing Zone architecture developed by Microsoft is available at `https://learn.microsoft.com/en-us/azure/cloud-adoption-framework/ready/landing-zone/#azure-landing-zone-architecture`.

- Leverage Azure Policy to enforce actions during the creation and deletion of the Azure service. For example, using Azure Policy, we can automatically apply tags to resources once they are created.

- Enforce consistency across resources using Azure Policy. For example, if you want to automatically install Azure Monitor Agent to all virtual machines, we can apply a policy for that. When the VM is created, Azure Policy will install the agent without any manual intervention.

- Know your regulatory requirements and apply the corresponding standards. For instance, if your organization requires HIPAA standards, we can apply initiative and verify compliance regularly. The following figure shows built-in regulatory compliance definitions offered by Azure:

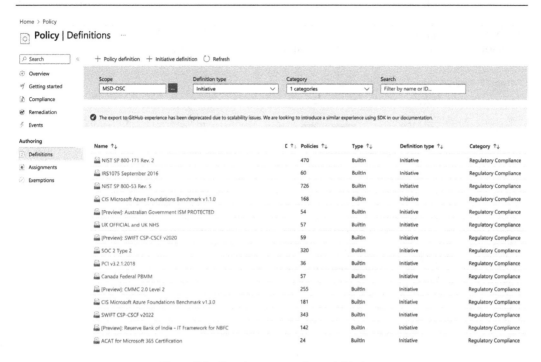

Figure 7.3 – Regulatory compliance initiatives

- Once the policies are assigned, verify compliance from the Azure portal or using Azure Resource Graph. Continuous evaluation is required for all applied policies. The following figure shows compliance with applied policies from the Azure portal; we can see compliance for individual policies including the count of compliant and non-compliant resources:

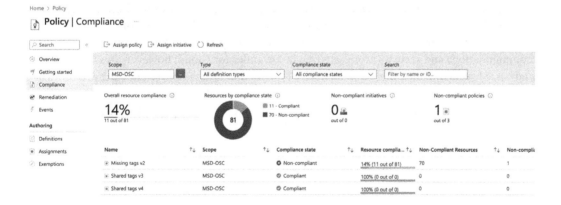

Figure 7.4 – Compliance dashboard

- Review and apply security recommendations available in Azure Advisor. The following figure shows how you can access security recommendations from Azure Advisor:

Figure 7.5 – Azure Advisor recommendations

- Perform vulnerability assessments to keep attackers away.

Now that we've discussed the governance checklist, let's move on to the Azure Security Benchmark.

### Azure Security Benchmark

The recommendations you saw in *Figure 7.6* are sourced to Azure Advisor from Microsoft Defender for Cloud. Using Microsoft Defender for Cloud, we can improve the security of your Azure and non-Azure environment, even if you don't have skilled security professionals in your environment. Recommendations and security scores are free in Defender for Cloud; however, you need to pay for advanced security suites offered by Defender for Cloud. It is advisable to onboard your critical workloads in Azure and non-Azure environments to Defender for Cloud. The following figure shows an overview of resources (across Azure, AWS, and GCP) and the security benchmarks:

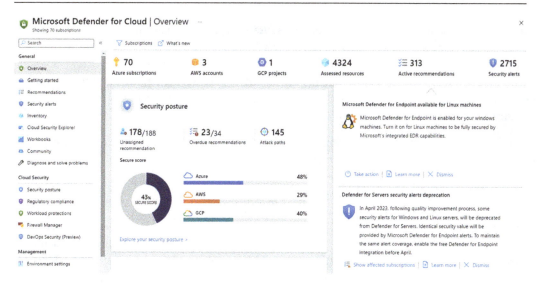

Figure 7.6 – Microsoft Defender for Cloud

You might be wondering how Microsoft Defender for Cloud is able to come up with the recommendations that you are seeing in Azure Advisor. Here comes the role of Azure Security Benchmark, which comprises a set of best practices defined by Microsoft that are part of industry standards, such as **Center for Internet Security (CIS)** controls, **National Institute of Standards and Technology (NIST)**, and the **Payment Card Industry Data Security Standard (PCI-DSS)** framework. With the help of Microsoft Defender for Cloud, we will be able to monitor the compliance status of our resources and verify whether they are aligned with Azure Security Benchmark recommendations. During Microsoft Defender for Cloud configuration, the Azure Security Benchmark initiative will be assigned to the subscription with the name `ASC Default (subscription: <subscription ID>):`

Figure 7.7 – Azure Security Benchmark compliance

You can see the full list of controls that are part of Azure Security Benchmark at `https://learn.microsoft.com/en-us/azure/governance/policy/samples/azure-security-benchmark`. The bottom line is if you want to immediately increase the security posture of your environment, you can use Microsoft Defender for Cloud, which uses Azure Security Benchmark to assess your environment and come up with recommendations and enhancements. Azure Security Benchmark is used in all design areas of security.

With that, we have completed the governance design principles; now, we will move on to the next design area: identity and access management.

## Identity and access management

Every day, we interact with identity and access management solutions – for example, when you are signing in to solutions such as email accounts, booking flight tickets, and mobile applications. Identity management is the process of authenticating and authorizing users, groups, and applications. Rather than developing and managing homegrown identity solutions, you should leverage **identity and access management** (**IAM**) solutions such as Azure AD. By doing so, you don't have to worry about managing the infrastructure or updates, as all of that will be taken care of by Microsoft. The following services are used for identity in Azure (links are added for your reference):

- Azure AD (`https://learn.microsoft.com/en-us/azure/active-directory/`)

- Azure AD B2B (`https://learn.microsoft.com/en-us/azure/active-directory/b2b/`)

- Azure AD B2C (`https://learn.microsoft.com/en-us/azure/active-directory-b2c/`)

Furthermore, we can bring identities from Active Directory to Azure with the help of the Azure AD Connect tool. Reference architecture to integrate on-premises Active Directory domains with Azure Active Directory is available at `https://learn.microsoft.com/en-us/azure/architecture/reference-architectures/identity/azure-ad`.

As we have seen in governance, we have a checklist in the case of IAM as well. The following factors should be on the checklist while managing identity for your workloads:

- While granting access, always follow the principle of least privilege. We covered this concept in the *Understanding the security pillar principles* section.

- With the help of RBAC, assign permissions to different scopes such as management groups, subscriptions, resource groups, or individual resources. Always use built-in roles, if possible; otherwise, use custom roles.

- Use resource locks at subscription, resource group, or resource scope to avoid accidental deletion or modification.

- Use managed identities of Azure resources to access other resources in Azure.

- Though we have the option to create multiple directories, it's better to take a single-directory approach.

- While synchronizing on-premises identities to the cloud, ensure that you are not synchronizing critical on-premises accounts to the cloud.

- Configure Azure AD Conditional Access to enforce policies that will be triggered when users authenticate. For example, you can allow access to applications from certain geographic regions only.

- Use separate identity sources for consumers and partners. You can use Azure AD B2B for employees and partners, while Azure B2C is used for all consumers.

- Leverage passwordless methods for authentication.

- Block legacy protocols (SMTP, IMAP, etc.) and authentication methods. You can use an Azure AD Conditional Access policy to enforce and block all legacy protocols and authentication methods.

Having explored the IAM design, we are now turning to the next topic, which is network security.

## Network security

We need to protect our resources by implementing **network security** controls by inspecting all traffic from and to Azure. In the absence of these security controls, attackers will be able to gain access to your resources. Ideally, we should implement more security on our external-facing workloads as attackers will be constantly scanning public IP ranges and looking for vulnerabilities.

The following Azure services can be used to implement defense-in-depth elements to detect, contain, and stop attackers from gaining access to your cloud environment:

- Azure Firewall

- Azure Virtual Network NAT

- Azure DDoS Protection

- Azure Web Application Firewall

Apart from native Azure Firewall, you can also deploy NVA solutions from Azure Marketplace. Popular vendors' appliances such as Palo Alto, Fortinet, and F5 are available in Azure. If you already have licenses from these vendors, you can use them in Azure as well. One key aspect to consider here is that Azure Firewall is a PaaS solution, while NVAs are deployed as IaaS VMs in Azure. Azure Firewall has built-in availability and scalability; however, in the case of NVAs, you need to deploy multiple VMs to make them highly available. The reference architecture to deploy highly available NVAs can be found at https://learn.microsoft.com/en-us/azure/architecture/reference-architectures/dmz/nva-ha.

Most organizations will be following the Enterprise Scale Landing Zone and adopt hub-spoke topology in Azure. The reference architecture of hub-spoke network topology in Azure is available at `https://learn.microsoft.com/en-us/azure/architecture/reference-architectures/hybrid-networking/hub-spoke`. In this architecture, you can replace Azure Firewall with highly available NVAs to use third-party network appliances. On that note, let's look at the checklist for network security:

- Segment your network and ensure that the communication between network segments is secured

- Ensure that you have security measures to filter network traffic, manage access requests, and communicate between application components

- All public endpoints should be protected using Azure Front Door, Azure Application Gateway, Azure Firewall, or Azure DDoS Protection (to protect public IP addresses, as recommended by Microsoft)

- A Basic DDoS Protection plan is enabled for all Azure virtual networks; however, it's recommended to enable a Standard plan for networks that are using public IP addresses

- Allow VMs to connect privately and securely to the internet using Azure Virtual Network NAT, also known as NAT Gateway

- Traffic between subnets (east-west) and application tiers (north-south) should be controlled using custom routes allowing the traffic to be inspected by Azure Firewall or NVA

- Defense-in-depth approach and controls should be applied at each level to protect from data exfiltration

With that, we are concluding the checklist for network security. As we continue our discussion of design areas, let's cover data protection next.

## Data protection

Sensitive data should be always handled with great care. Misconfiguration of data sources may lead to data leaks. We need to use access control, encryption, and logging in Azure to classify, protect, and monitor sensitive data assets in Azure. Azure Security Benchmark, which we discussed in the *Governance* section, has policies that focus on data protection. We will start with a checklist for data protection:

- Use Azure AD-based authentication for supported Storage services. In this case, we don't have to manage access keys or **shared access signatures** (**SAS**) to access storage; authorization is managed with the help of Azure RBAC. Use Azure AD with databases such as SQL Database to secure your databases.

- Instead of developing homegrown encryption, leverage native encryption offered by Azure services.

- All data in storage should be encrypted and should have data classification.

- Always encrypt data in transit to avoid any unauthorized access.

- Access keys and connection strings should be stored in Azure Key Vault and access to Key Vault should be managed using access policies.

- If you are using storage access keys to access Azure Storage, make sure that you rotate the keys frequently.

- Like keys, all secrets should be regenerated periodically.

While the checklist was an important topic to cover, the next topic (storage encryption) will allow us to expand our understanding even further.

### Storage encryption

In a solution, a single operation may lead to multiple data transactions, and as part of the process, data movement will occur. In order to implement data protection, we need to encrypt the data while it is in storage and while it is in transit. When it comes to data protection, we need to encrypt the data based on the state of the data:

- **Encryption at rest**: All data that is stored in physical media (magnetic or optical disk) is called data at rest, and we need to encrypt this data. If an intruder manages to steal the disk, they will not be able to read the data on the disk without the decryption keys.

- **Encryption in transit**: Data that is being transferred between application components or locations is called data in transit. By encrypting data in transit, we ensure that unauthorized access is restricted.

Besides the checklist for data protection that we discussed earlier in this section, the following key points should be considered when it comes to storage encryption:

- Do not develop your own encryption standards; use standard and recommended encryption algorithms

- Secure hash algorithms such as the SHA-2 family should be used

- Enable data classification on top of data at rest and have separate encryption based on the classification of data

- Leverage Azure Disk Encryption to encrypt the virtual disks of your VMs

- Secure your **data encryption key** (**DEK**) with the help of a **key encryption key** (**KEK**)

- Use TLS 1.2 on Azure

- Data in transit for client/server communication should be encrypted over TLS/HTTPS network channels only

Now that we have a solid understanding of storage encryption, let's build upon that foundation by examining key and secret management.

### *Key and secret management*

As we discussed in the previous section, we need to encrypt the storage using keys. Now the question would be how to secure our keys and secrets. As a side note, when you see *storage*, it is not only referring to Azure Storage, it also includes data stores such as databases as well. Databases will have connection strings that will be used by our application to read and write data. These connection strings need to be stored in a secure way instead of writing them as plain text in your code. Though we have discussed the following points in the checklist, let's recap the key considerations for key and secret management:

- Identity-based access control should be used in place of cryptographic keys

- Azure Key Vault or a similar key management service should be used to store keys and secrets securely

- Periodically rotate your secrets and keys

If you are using the following keys and secrets in your environment, then you should secure them to avoid security leaks:

- API keys

- Database connection strings

- **DEKs**

- **KEKs**

- Passwords

You should never handle keys and secrets in your code or configuration. If an attacker manages to gain access to your code, then they will be able to hijack your keys and secrets. The recommendation is to secure keys and secrets in a key management solution.

In Azure, we have Azure Key Vault, which can help you store keys and secrets securely. Other services such as HashiCorp Vault can be leveraged if you are already using them. In the case of Azure Key Vault, we can use Azure RBAC and access control policies to provide authorization to access keys and secrets. You can create managed identities to access the keys and secrets in Azure Key Vault securely.

As we transition to applications and services, it's important to consider the impact of key and secret management in your security posture.

## Applications and services

In the previous sections on design areas, we covered governance, IAM, network security, and data protection; now we will move on to the last design area: **applications and services**. We will divide this section into three broad topics, as follows:

- Classifications

- Securing PaaS deployments
- Configuration and dependencies

We will start with application classification.

### Application classification

Organizations typically have large application portfolios, but not all of them are critical. We need to classify our applications based on criticality. For instance, business-critical applications should have tightened security measures. If no proper security measures are applied, the chances of security failures are very high. Compromised applications lead to revenue and reputation loss. Identify and classify your applications based on the data type and data criticality. The following are some examples of classification:

- **Business-critical data**: Applications that handle business-critical data must have confidentiality, integrity, and availability assurance
- **Regulated data**: Applications that handle PII information, health data, or payment instruments should have compliance standards such as PCI-DSS, HIPAA, and so on

You should always prioritize applications based on their criticality and apply the necessary security controls. You can review the criticality scale (`https://learn.microsoft.com/en-us/azure/cloud-adoption-framework/manage/considerations/criticality#criticality-scale`), which is part of the Cloud Adoption Framework, to classify your applications. On that note, we will move on to securing PaaS deployments.

### Securing PaaS deployments

PaaS services offer a lot of security advantages when compared to IaaS deployments. If you remember the shared responsibility model, most of the responsibilities will be taken care of by Microsoft in the case of PaaS solutions. The security advantages of PaaS can be viewed at `https://learn.microsoft.com/en-us/azure/security/fundamentals/paas-deployments#security-advantages-of-a-paas-cloud-service-model`.

Microsoft's cloud infrastructure is secure and hard to attack due to continuous monitoring. Attackers are unlikely to target the Microsoft cloud unless they have significant resources. PaaS deployment and on-premises have similar risks at the application and access management layer, and best practices can help minimize these risks. At the top of the stack, key management can mitigate the risk of data governance and rights management. Azure provides strong DDoS protection using network-based technologies, but there are limits. To defend against large DDoS attacks, Azure allows for quick and automatic scaling.

Now we will move on to configuration and dependencies.

### Configuration and dependencies

The safeguarding of an application that is hosted on Azure is a joint responsibility shared between the application owner (you) and Azure. If you're using IaaS, you're accountable for configurations that concern the VM, operating system, and components that are installed on it. If you're using PaaS, you'll be responsible for the security of your application and service configurations as well as ensuring that the dependencies used by your application are also secure.

The following key points should be considered for the security of configuration and dependencies:

- Don't store secrets in your code and configuration files. Instead, store them in secure storage such as Azure Key Vault or Azure App Configuration.

- In the case of handling exceptions, don't share detailed error data.

- Don't reveal platform-specific information.

- Application configuration should be stored outside the application code and should be updated separately.

- If a resource is not meeting security requirements, don't let it access your application.

- Any open source code used should be validated before using in production.

- As part of the application life cycle, update frameworks and libraries periodically.

With that, we can conclude our discussion on design areas and move on to the last topic of this chapter – monitoring.

# Monitoring

Regularly monitoring resources is crucial to maintaining a strong security posture and detecting vulnerabilities. This detection can occur in the form of proactively searching for anomalous events within enterprise activity logs or reacting to alerts of suspicious activity. It is important to respond to any anomalies or alerts in a prompt and vigilant manner in order to prevent any reduction in security assurance. Additionally, employing defense-in-depth and least privilege strategies are key to designing a strong and secure system.

The following checklist should be used for monitoring security-related events in this workload:

- Use Azure Monitor to monitor workloads deployed on Azure

- As part of the incident response plan, invest in building a **Security Operations Center** (**SOC**) or SecOps team

- Traffic from or to applications, access requests, and application communication should be monitored

- Leverage the secure score in Microsoft Defender for Cloud to review and remediate recommendations

- Use Azure Security Benchmark and other industry standard benchmarks to assess the security posture of your environment

- Send logs and alerts to SIEM tools such as Microsoft Sentinel for analysis and log management

- Conduct periodic internal and external audits to verify compliance

- Perform penetration testing and security evaluation to capture flaws in the architecture

You can use the following reference architectures to improve the security landscape of your environment:

- **Hybrid security monitoring using Microsoft Defender for Cloud and Microsoft Sentinel**: Combine Microsoft Defender for Cloud and Microsoft Sentinel to monitor security configuration and telemetry from on-premises and Azure workloads. You can take a look at the reference architecture at `https://learn.microsoft.com/en-us/azure/architecture/hybrid/hybrid-security-monitoring`.

- **Azure security solutions for AWS**: Azure security solutions such as Microsoft Defender for Cloud and Microsoft Sentinel not only cover Azure and on-premises workloads but they can also be used to monitor your workloads deployed in **Amazon Web Services** (**AWS**) and **Google Cloud Platform** (**GCP**). To better understand how Azure security solutions work with AWS, you can refer to `https://learn.microsoft.com/en-us/azure/architecture/guide/aws/aws-azure-security-solutions`.

With that, we've covered all the major points relevant to the security pillar. Before we move on to the next chapter, let's briefly review the key takeaways from this one.

## Summary

In this chapter, we covered the last pillar of the WAF: security. We already covered other pillars such as cost optimization, operational excellence, performance efficiency, and reliability in the previous chapters. We started the chapter with an introduction to security and the relevance of security controls. Then we discussed the key areas and security resources. These are the core topics of this chapter; the later sections were mere expansions of the ideas covered in the key areas and security resources. After covering the key areas, we covered design areas such as governance, identity and access management, network security, data protection, and application and services. The last section in this chapter was about monitoring, where we covered a checklist for monitoring and reference architectures. Monitoring workload security can help administrators to irradicate attacks before they reach your data. Notifications and dashboards should be set up so that SOCs can identify potential risks and mitigate them in a holistic fashion.

So far, we have been focusing on the design principles for the pillars of the WAF and never got a chance to assess workloads. In the next chapter, you will get a chance to review workloads and come up with recommendations based on the pillars of the WAF. As we transition to the last chapter of this book, let's shift our focus to assessment and remediation and the exciting new ideas we'll encounter there.

# Part 3: Assessment and Recommendations

Once we are familiar with the concepts and design areas of the Well-Architected Framework, it's crucial that we assess our existing workloads and ensure that they align with the Well-Architected Framework principles. This part focuses on assessing and remediating workloads with the help of Microsoft Assessments and a reference workload. The reference workload will be used as an example, and we will come up with remediations that can potentially optimize the workload to make it cost optimized, reliable, secure, and efficient. This part has only one chapter:

- *Chapter 8, Assessment and Remediation*

# Assessment and Remediation

To those who have turned the pages with unwavering dedication, thank you for investing your time and energy in this book. Congratulations, you have successfully completed all pillars of the **Well-Architected Framework (WAF)**. So far, we have only focused on the theory and design aspects of the WAF pillars. In this chapter, we will cover the WAF questionnaire shared by Microsoft and furthermore, we will review a reference architecture and see how we can align it to the WAF pillars, namely cost optimization, operational excellence, performance efficiency, reliability, and security.

The review is conducted using the **Microsoft Assessments Portal** and we will have a three-tier application as our reference workload. The optimizations and recommendations will be derived for the reference workload, which will be further used to align our workload to the WAF principles. As described, two key tools you need to understand in this chapter are the **assessment tool** and **reference workload**. We will start by discussing the assessment tool.

## Introducing the assessment tool

Depending on the pillar you want to assess, we use separate questionnaires. The initial steps to start the assessment are the same for all pillars, so we will first familiarize you with the assessment interface.

You can review the questionnaire at `https://learn.microsoft.com/en-us/assessments/azure-architecture-review`. To access it, follow these steps:

1. Navigate to the link, and you will see the **Start Assessment** button, as shown in the following figure:

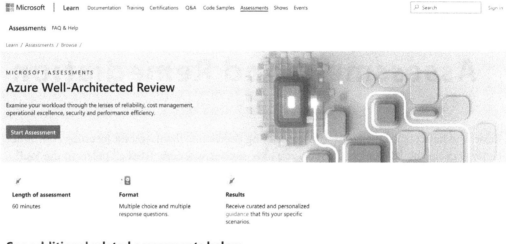

Figure 8.1 – Azure Well-Architected Review

If you look closely at the preceding screenshot, you can see other additional related assessments such as **Azure Landing Zone Review**, **DevOps Capability Assessment**, and **Data Services | Well-Architected Review**. Depending on your organizational requirement, you can conduct other assessments from this portal. Coming back to **Azure Well-Architected Review**, once you click on **Start Assessment**, you will be asked to sign in, as shown in the following screenshot:

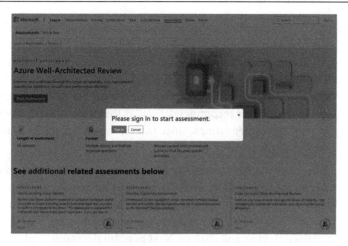

Figure 8.2 – Sign-in prompt to start the assessment

2. You can sign in using your Microsoft Learn account. In this way, all your assessments will be stored in your account, and you can review these later whenever you want.

   If you don't have a Microsoft Learn account, you will be asked to complete the sign-up process.

3. After signing in, you will be presented with a screen similar to the one shown in *Figure 8.1*, and you need to click on **Start Assessment** to start the assessment with your signed-in account.

4. In the next screen, you will be asked to give a name for the assessment so that you can distinguish it from other assessments that you will conduct later. The recommendation is to give a meaningful name along with the date so that you can easily recognize the assessment:

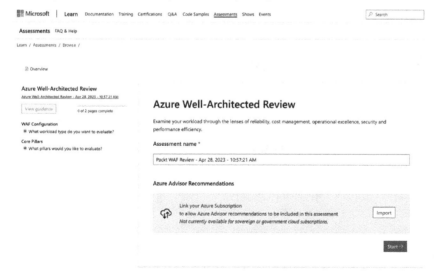

Figure 8.3 – Starting Azure Well-Architected Review

5. You can link your Azure subscription using the **Import** button. Using the import feature, you can bring in your **Azure Advisor** recommendations to the final report if you need a comprehensive assessment report. This will ensure that your report contains recommendations based on your responses and recommendations from Azure Advisor as well. Otherwise, the final report will be based only on the responses that you give to the questions. For this demonstration, we will not be importing Azure Advisor recommendations as the demo subscriptions we are using don't have any Advisor recommendations. Once you are done, click on the **Start** button (refer to *Figure 8.3*) to initialize the review.

6. Once you start the review, you will be asked to select the type of workload you want to evaluate. We have different options such as **Core Well-Architected Review**, **Azure Machine Learning**, **Internet of Things**, **SAP On Azure**, and **Azure Stack Hub**. Our interest is in the **Core Well-Architected Review**, which will help us to assess our workloads using the WAF pillars. Select **Core Well-Architected Review** and click on **Next** to begin the assessment:

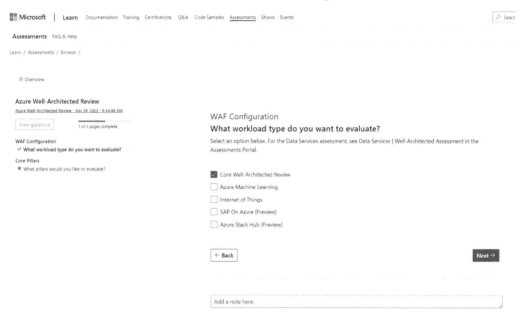

Figure 8.4 – Selecting the workload type

7.   In the next screen, you will be asked to select the pillar you would like to evaluate:

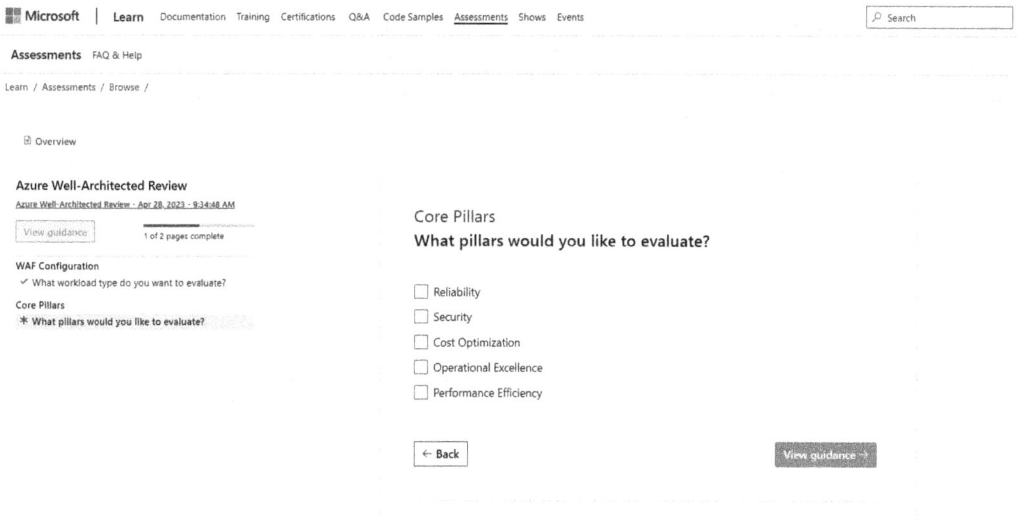

Figure 8.5 – Selecting the pillars

You can select all the pillars in a single step, or you can go one by one. In this chapter, we will be evaluating the pillars one by one rather than combining the questionnaires.

Now, we will pause, and when we reach the relevant pillar, the next steps will be discussed. These initial steps will be the same for all WAF reviews and further steps will be covered later in this chapter. Before that, we will take a look at the reference workload.

# Introducing the reference workload

We will be using a Windows N-tier application on Azure architecture (`https://learn.microsoft.com/en-us/azure/architecture/reference-architectures/n-tier/n-tier-sql-server`) from the Azure Architecture Center as our reference architecture. Various aspects such as networking, load balancing, SKUs used, and performance metrics will be described in this section. We will be using this data to verify the alignment with the principles of the WAF. When you conduct an assessment for your workloads, you can choose any of your solutions and ensure that it aligns with the WAF pillars.

As we continue our discussion of the reference workload, let's take a look at the architecture.

## Architecture

Though we are using the N-tier application reference architecture, as you can see in *Figure 8.5*, there are some minor differences, such as the Active Directory Domain Services subnet has been removed from the diagram. Nevertheless, the idea is to make you understand how we can work with the pillars and principles of the WAF regardless of the type of workload you have. You can take an in-depth look at this at `https://learn.microsoft.com/en-us/azure/architecture/icons/#icon-terms`.

Shifting gears, let's now take a look at the workflow of the architecture.

## Workflow

The application has variable traffic comprising low and high usage. As of now, the architecture was taken from the Azure Architecture Center and we are using an altered version of that. We need to ensure that cost optimization, performance efficiency, operational excellence, reliability, and security assessments are done on this workload. In this section, we also discuss resource-related information such as SKUs and performance metrics (wherever applicable), which can be quite useful in developing the recommendations. The data will be represented in the form of tables for easy representation and understanding.

The architecture comprises the following components:

- **Resource group**: All our application tiers are mapped to a single resource group for logical grouping. Furthermore, we can also manage compliance, governance, access, and the life cycle of resources from the resource group scope.

- **Availability zones**: A physical location within an Azure region is called an availability zone. Within an availability zone, we have zones that represent a group of data centers. These zones are separated from each other and are connected using low-latency optical fiber. Each zone is equipped with independent power, cooling, and networking components to avoid a single point of failure. In this architecture, VMs are placed in different zones to improve the reliability of the application.

- **Virtual network and subnets**: We have a single virtual network with five subnets. The Application Gateway subnet is used by Application Gateway and the Management subnet is used by jump box servers. The remaining three subnets are used by application tiers, namely the web tier, business tier, and data tier.

- **Application Gateway**: This provides Layer 7 load balancing for HTTP/HTTPS requests. All web requests coming to our application will be routed by Application Gateway to the servers in the web tier. In addition to load balancing, Application Gateway also offers Azure Web Application Firewall, which protects the application from common web vulnerabilities. The specifications of Application Gateway are shown in the following table:

| Instance Count | SKU | Performance Metrics |
| --- | --- | --- |
| 2 | Standard | Throughput: 14,000 bytes/sec |

Table 8.1 – Application Gateway specifications

- **Azure Load Balancer**: This provides Layer 4 load balancing. All requests from the web tier will be load balanced by the Load Balancer to the business tier servers. Similarly, requests from business tier servers will be load balanced to the database servers using Azure Load Balancer. The specifications of Azure Load Balancer are shown in the following table:

| Tier | SKU | Performance Metrics |
| --- | --- | --- |
| Business tier | Standard | Byte count: 14.2K |
| Data tier | Standard | Byte count: 10.7K |

Table 8.2 – Azure Load Balancer specifications

- **Network Security Groups** (**NSGs**): In this architecture, we are using NSGs to filter the network traffic. For example, we are explicitly denying direct traffic from the web tier to the data tier. If the web tier needs to collect data from the data tier, it needs to go through the business tier. If you are using Azure Landing Zone and a hub-spoke architecture, you can route the requests to Azure Firewall or **Network Virtual Appliance** (**NVA**) before they are routed to the next tier with the help of user-defined routes.

- **DDoS Protection**: Since we have external-facing public IPs, it's better to upgrade from the Basic DDoS Protection plan to the Standard plan, which offers advanced DDoS mitigation features.

- **Azure DNS**: In this architecture, Azure DNS provides name resolution by hosting domains in Azure.

- **Virtual machines**: We have Windows virtual machines running all tiers of the application. In the case of the data tier, we have the SQL Server Always On availability group feature enabled for the high availability of the data tier. Furthermore, we have jump box servers used for management purposes with public access. Specifications for the virtual machines are as follows:

| Server Role | Count | SKU | Performance Metrics |
| --- | --- | --- | --- |
| Web tier servers | 3 | D16Sv3 | CPU: 8%, Memory: 11% |
| Business tier servers | 3 | D8Sv3 | CPU: 4%, Memory: 6% |
| Data tier servers | 2 | E4Sv3 | CPU: 52%, Memory: 49% |
| Jump box servers | 1 | D8Sv3 | CPU: 13%, Memory: 12% |

Table 8.3 – Virtual machines' specifications

- **Cloud witness**: Since we are setting up an SQL Always On availability group with the help of **Windows Server Failover Cluster** (**WSFC**) technology, the failover cluster always requires more than half of the nodes running, thus establishing a quorum. Sometimes, if the cluster has only two nodes running, this could lead to a tie, where each node assumes that it's the primary node. In order to break this tie and establish a quorum, we use Azure Blob Storage as a type of witness.

Having discussed both assessment tools and the reference workload, now we will start with the cost optimization assessment and its remediation.

## Cost optimization assessment and remediation

In this section, we will see how we can use the questionnaire for cost optimization in the Assessments portal and perform workload optimization of our reference architecture. As discussed in *Chapter 3, Implementing Cost Optimization*, teams need to have a shift in their mindsets. Ideas that you learned from on-premises should be remolded before applying to the cloud due to the differences in environment. For example, in on-premises, we can host a website in a VM or a container, but in the cloud, we have different IaaS and PaaS offerings based on our requirements. Always take an open-minded approach so that you can learn about the cloud better and get the best out of it. With that in mind, let's now turn to the Assessments portal and see the questionnaire for cost optimization.

### Questionnaire

Now that we are going to take the cost optimization questionnaire, we need to circle back to *Figure 8.5*, where we saw all the pillars of the WAF presented to us to review the guidance. Since we need to perform a cost optimization assessment, you can select **Cost Optimization** and click on **Next**, as shown in the following screenshot:

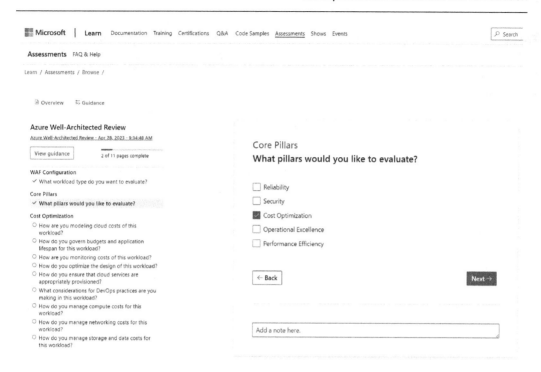

Figure 8.6 – Evaluating the Cost Optimization pillar

As you can see, the moment you select **Cost Optimization**, questions will be populated on the left-hand side of the screen.

The questionnaire consists of the following questions:

- **How are you modeling cloud costs of this workload?**

- **How do you govern budgets and application lifespan for this workload?**

- **How are you monitoring costs of this workload?**

- **How do you optimize the design of this workload?**

- **How do you ensure that cloud services are appropriately provisioned?**

- **What considerations for DevOps practices are you making in this workload?**

- **How do you manage compute costs for this workload?**

- **How do you manage networking costs for this workload?**

- **How do you manage storage and data costs for this workload?**

For each question, there will be detailed descriptions and embedded videos explaining the concept, and the relevance of the question. Based on your current organizational setup, you need to answer these questions. Once you complete the questionnaire, you will be presented with a score and an improvement plan:

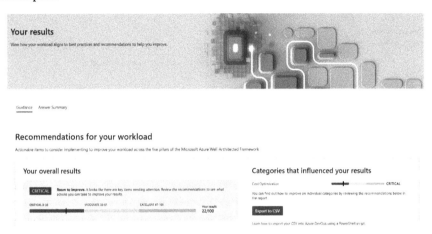

Figure 8.7 – Cost optimization result

As per the inputs given, the tool concluded that the environment is in a critical stage and requires attention. The overall score achieved is **22/100**, which explains why it's critical. You can download the key points from individual categories to a CSV file for your analysis by clicking on the **Export to CSV** button. Furthermore, on the same page, recommendations will be outlined in a table with their priority levels, as shown in the following figure:

Figure 8.8 – Recommendations

If you have linked your Azure subscription before starting the assessment, Azure Advisor recommendations related to cost will be automatically included in the recommendation list. Teams can export the list to project management tools to track the activities and finish them one by one based on priority.

The process is the same for other pillars of the WAF; nevertheless, we will review these questions in the respective sections. As we continue, let's examine and try to derive recommendations for your reference workload.

## Reference workload

In this section, let's see what cost-related recommendations can be derived for the application based on the data we have. Let's consider the services one by one and come up with recommendations and a remediation plan:

- **Virtual machines**:

| Topic | Recommendation | Remediation Plan |
|---|---|---|
| Right-sizing based on utilization | Web servers and business tier servers are underutilized; try resizing them to a smaller size. | Review Azure Advisor to see the resizing plan and quick-fix measures. |
| Migrate to **Virtual Machine Scale Set (VMSS)** | Instead of going for a fixed number of instances, consider moving the web tier and business tier to VMSS for autoscaling. | After resizing, if the utilization is not optimized, consider switching to VMSS to automatically manage the number of instances. |
| Azure Hybrid Benefit | Consider getting Azure Hybrid Benefit for Windows VMs if you have already purchased licenses. | Enable Azure Hybrid Benefit on VMs to save cost. |
| Reserved instances | If you plan to keep the application for a longer period of time, try reserving instances for 1 or 3 years. | Review reservation purchase recommendations in Azure Advisor and purchase accordingly. |
| Azure Hybrid Benefits for SQL | Consider enabling Azure Hybrid Benefit for SQL VMs if you have already purchased licenses. | Enable Azure Hybrid Benefit on SQL licensing to save cost. |

| SQL migration to PaaS | Evaluate the price difference between IaaS VMs and PaaS databases. PaaS offers better availability and management. | If feasible, use database migration tools to move from SQL VM to SQL PaaS. |

Table 8.4 – VM recommendations

- **Application Gateway**:

| Topic | Recommendation | Remediation Plan |
| --- | --- | --- |
| Application Gateway autoscaling | Application Gateway is offered in fixed and autoscaling models. Use autoscaling to save cost. | Upgrade to Standard v2 to use the autoscaling feature. |

Table 8.5 – Application Gateway recommendations

- **Load Balancer**:

| Topic | Recommendation | Remediation Plan |
| --- | --- | --- |
| Load Balancer rules | Verify the number of rules for load balancing and outbound rules as these are chargeable. | Remove unwanted rules. Inbound NAT rules are free of cost. |

Table 8.6 – Load Balancer recommendations

In this way, you can optimize your workloads and save costs. The number of recommendations will vary from workload to workload. Most of the recommendations such as reserved instances, right-sizing, and so on are available in Azure Advisor, and we can import these to the assessment tool by linking our Azure subscription.

With that, we have completed the cost optimization assessment; now, we will move on to assessing the next pillar – operational excellence.

# Operational excellence assessment and remediation

Similar to cost optimization, we will start with the questionnaire and then we will analyze our reference application architecture. We can start the assessment again with a new name and select the **Operational Excellence** pillar to conduct the review. The review will have questions that will be used to evaluate your alignment with the principles of operational excellence. Let's start with the questionnaire.

## Questionnaire

Assuming that you have already completed the initial steps outlined in the assessment tool section, you will be required to select **Operational Excellence** and click on **Next** to start the evaluation:

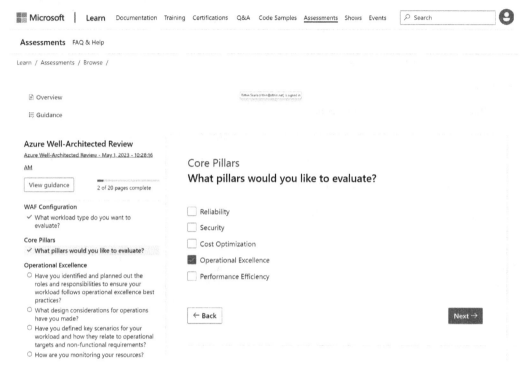

Figure 8.9 – Starting the Operational Excellence evaluation

The following questions are part of the evaluation:

- **Have you identified and planned out the roles and responsibilities to ensure your workload follows operational excellence best practices?**

- **What design considerations for operations have you made?**

- **Have you defined key scenarios for your workload and how they relate to operational targets and non-functional requirements?**

- **How are you monitoring your resources?**

- **How do you interpret the collected data to inform about application health?**

- **How do you visualize workload data and then alert relevant teams when issues occur?**

- **How are you using Azure platform notifications and updates?**

- **What is your approach to recovery and failover?**

- How are scale operations performed?

- How are you managing the configuration of your workload?

- What operational considerations are you making regarding the deployment of your workload?

- What operational considerations are you making regarding the deployment of your infrastructure?

- How are you managing and distributing your patches

- How are you testing and validating your workload?

- What processes and procedures have you adopted to optimize workload operability?

- What operational excellence allowances for reliability have you made?

- What operational excellence allowances for cost have you made?

- What operational excellence allowances for security have you made?

Once the evaluation is done, you will be presented with the overall results, as shown in the following screenshot:

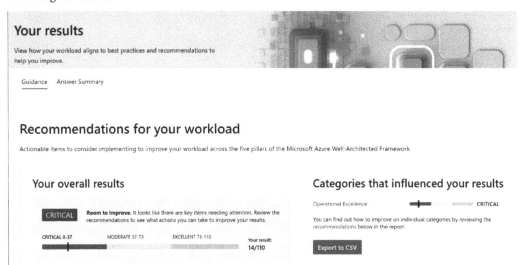

Figure 8.10 – Operational Excellence evaluation results

Furthermore, you can review the recommendations in the **Improve your results** section:

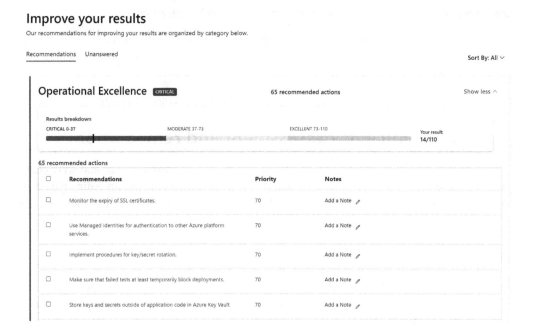

Figure 8.11 – Operation Excellence evaluation recommendations

With that, we have finished the questionnaire; now we will review the reference workload and try to come up with recommendations.

## Reference workload

Having all application tiers as part of a single virtual network within a single resource group enables easier management of resources. Policies, access management, locks, and so on can be managed at the resource group level. Furthermore, continuous integration and continuous delivery can be easily implemented for the workload. For the virtual machines, we can use the virtual machine extensions for post-deployment configurations. With the help of extensions, we can automatically install software and achieve the desired state without manual intervention. We have already covered extensions such as **Desired State Configuration (DSC)** and Custom Script Extension earlier in *Chapter 4, Achieving Operational Excellence*. Since we are maintaining an SQL Always On availability group, we can use the DSC extension to make sure that the nodes are correctly configured.

Operational excellence is all about setting up automation and collecting monitoring data. When it comes to monitoring, we can onboard our resources to Azure Monitor. We can use metrics and logs collected from the resources for troubleshooting and to detect issues before they turn into bigger issues.

In addition to the collection of logs and metrics, we need to onboard our application to **Application Insights**, which will help us get logs from the application stack, including live metrics. In short, we need to use Azure Monitor to monitor the infrastructure and application.

Since we are integrating our application into DevOps, we can include testing whenever a new code is published. We will be storing the code in source control and when a new commit is made, testing mechanisms are triggered to ensure the sanity of the code.

As we continue, let's move on to the next pillar in the WAF – performance efficiency.

## Performance efficiency assessment and remediation

With performance efficiency, we target the workload to withstand varying demands and serve client requests. Here also, we will take the same approach that we have taken in the two preceding pillars, so let's start with the questionnaire and review the architecture. Like the other pillars, the questionnaire completely focuses on improving the performance efficiency of your environment.

### Questionnaire

As of now, you might be familiar with the steps to start the assessment and select the pillar you want to evaluate. As shown in the following screenshot, you have to select the **Performance Efficiency** pillar to start the review process:

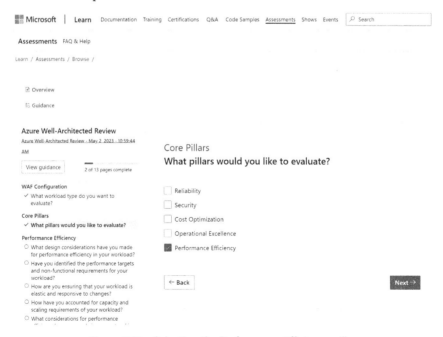

Figure 8.12 – Selecting the Performance Efficiency pillar

The following questions will be asked during the review process:

- **What design considerations have you made for performance efficiency in your workload?**
- **Have you identified the performance targets and non-functional requirements for your workload?**
- **How are you ensuring that your workload is elastic and responsive to changes?**
- **How have you accounted for capacity and scaling requirements of your workload?**
- **What considerations for performance efficiency have you made in your networking stack?**
- **How are you managing your data to handle scale?**
- **How are you testing to ensure that your workload can appropriately handle user load?**
- **How are you benchmarking your workload?**
- **How have you modeled the health of your workload?**
- **How are you monitoring to ensure the workload is scaling appropriately?**
- **What common problems do you have steps to troubleshoot in your operations playbook?**

Once the review is completed, you will be presented with the overall score:

Figure 8.13 – Overall score

You will also be shown recommendations based on your score:

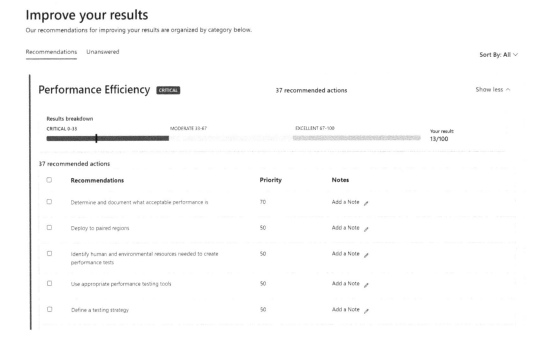

Figure 8.14 – Recommendations

Answer the questions based on your environment and let the system generate the recommendations for you. Based on the priority, you can complete the remediation. On that note, let's review the reference workload next and see what recommendations we can come up with.

## Reference workload

In the *Cost optimization assessment and remediation* section, one of the recommendations was to move from individual instances to a scale set. In fact, this is the same recommendation we have in terms of performance efficiency. With the help of scalability, we can make sure new instances are automatically added when there is an increase in demand. Similarly, when the demand is low, the instances are automatically deleted. If you are using individual instances, then you need to add the instance to the backend pool of the load balancer manually every time it's created; however, if you switch to the scale set, the mapping is done to the scale set. All new instances in the scale set will be automatically added to the load balancer for routing.

Another performance efficiency recommendation would be to use a cache instead of directly querying from the database. The cache will store frequently queried data and the application can instantly retrieve it without the need to reach out to the database.

If you are using Application Gateway, it's better to use autoscaling so that, based on the number of requests, the number of Application Gateway instances is automatically increased. As mentioned in cost optimization, this strategy can help in reducing the cost without compromising performance. Autoscaling is available in the Standard v2 SKU of Application Gateway.

As mentioned in the *Operational excellence assessment and remediation* section, integration with DevOps can help you conduct load testing, performance testing, and stress testing. With the help of these tests, we can design infrastructure with the surety that the infrastructure can manage the load. Furthermore, Azure Monitor is necessary for performance monitoring. With the help of Azure Monitor, we can track performance bottlenecks and mitigate them.

With that, we have completed the performance efficiency assessment and now we will delve into the assessment and remediation for the reliability pillar.

## Reliability assessment and remediation

We are nearing the end of this chapter, as Reliability is the second last pillar of the WAF we are going to discuss. Reliability is all about maintaining high availability of your workload such that you can meet metrics such as SLA, RPO, RTO, and so on. Our aim is to make sure that the services are up and running in case of catastrophic failure. In *Chapter 6, Building Reliable Applications*, we saw different techniques by which we can improve the high availability of our workloads. Let's look at whether our architecture is aligned with the best practices and techniques. We'll start with the questionnaire and get familiarized with the questions that are part of the reliability assessment.

## Questionnaire

By now, you know what to do if you want to conduct the Azure Well-Architected Review. We will start the assessment by selecting the **Reliability** pillar:

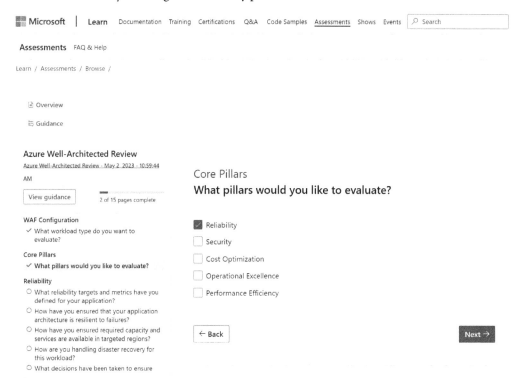

Figure 8.15 – Selecting the Reliability pillar

The following questions are part of the review:

- **What reliability targets and metrics have you defined for your application?**

- **How have you ensured that your application architecture is resilient to failures?**

- **How have you ensured required capacity and services are available in targeted regions?**

- **How are you handling disaster recovery for this workload?**

- **What decisions have been taken to ensure the application platform meets your reliability requirements?**

- What decisions have been taken to ensure the data platform meets your reliability requirements?

- How does your application logic handle exceptions and errors?

- What decisions have been taken to ensure networking and connectivity meets your reliability requirements?

- What reliability allowances for scalability and performance have you made?

- What reliability allowances for security have you made?

- What reliability allowances for operations have you made?

- How do you test the application to ensure it is fault tolerant?

- How do you monitor and measure application health?

Once the assessment is completed, you will be presented with the overall results, as shown in the following screenshot:

## Recommendations for your workload

Actionable items to consider implementing to improve your workload across the five pillars of the Microsoft Azure Well-Architected Framework

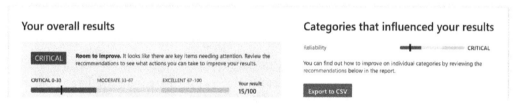

Figure 8.16 – Recommendations for the Reliability assessment

Furthermore, if you scroll down on the same page, you will be able to see individual recommendations including their priority level:

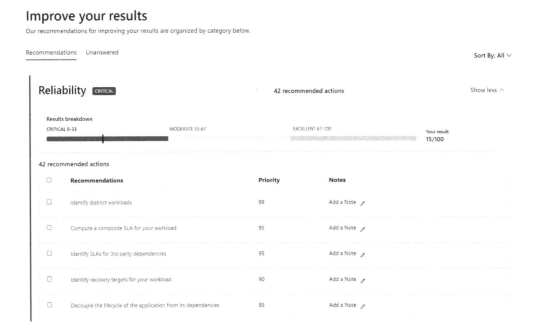

Figure 8.17 – Recommendations for improving reliability

Next, let's explore reference architecture-related recommendations.

## Reference workload

When it comes to the reliability of our reference architecture, we need to ensure that the services are highly available. For availability, you can consider the following recommendations:

- As of now, VMs are deployed across multiple availability zones, making the workload immune to zone-level issues. Having said that, if you need more availability, you should consider multi-region deployment. The reference architecture is shown at https://learn.microsoft.com/en-us/azure/architecture/reference-architectures/n-tier/multi-region-sql-server.

- If you consider the recommendation to migrate to VMSS, as mentioned in the *Cost optimization assessment and remediation* and *Performance efficiency assessment and remediation* sections, then you need to make sure that VMSS is deployed as a zone redundant service. For multi-region, you can deploy VMSS across multiple regions and load balance the requests using a load balancer service such as Traffic Manager.

- Set up health probes on the load balancer, which includes Azure Load Balancer and Azure Application Gateway. Health probes will validate whether the backend server is healthy to receive requests before routing the requests. In this way, if a server is down, the load balancer will stop routing the requests to the unhealthy server.

- Create an Azure Load Balancer and Azure Application Gateway as a zone-redundant service so that a load balancer is available in case of a zone failure.

- Though autoscaling Application Gateway is a performance efficiency-based recommendation, enabling autoscaling will help maintain the high availability of Application Gateway.

- If you are considering moving the SQL server on VM to a PaaS solution, setting up geo-redundancy is an easy task. However, if you plan to keep the SQL server on VM with multi-region deployment, then you need to set up asynchronous replication between the regions.

- Ensure Azure Monitor is set up to check the availability of services, and in the case of issues, administrators should be notified.

- Subscribe to Service Health and Resource Health alerts so that you are aware of planned maintenance and platform issues.

- With a single region approach for VMs, set up **disaster recovery** (**DR**) using Azure Site Recovery, which will help you to deploy the VMs to the DR region if the primary region is unavailable.

With the help of Azure DevOps, conduct reliability tests to make sure the application is reliable and is able to meet the organizational requirements and metrics.

Having examined the reference architecture recommendations, let's now shift our focus to the last pillar – security.

## Security assessment and remediation

Building secure applications can help you protect your application from attackers and security exploits. Day by day, attackers are coming up with more sophisticated attacks that will compromise your environment if security measures are not configured properly. Our focus is to build a defense-in-depth strategy and make it harder for attackers to penetrate your environment. As usual, we will start with the questionnaire and then move on to the reference architecture.

## Questionnaire

In the **Assessments** portal (`https://learn.microsoft.com/en-us/assessments/azure-architecture-review/`), when you start the assessment, you can select **Security** to start the security review:

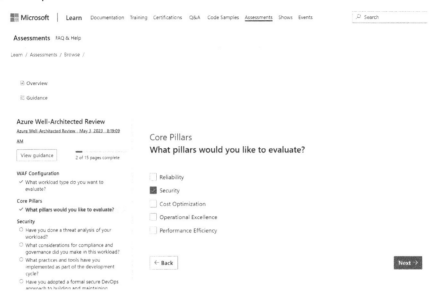

Figure 8.18 – Selecting the Security pillar

The following questions are part of the review:

- **Have you done a threat analysis of your workload?**

- **What considerations for compliance and governance did you make in this workload?**

- **What practices and tools have you implemented as part of the development cycle?**

- **Have you adopted a formal secure DevOps approach to building and maintaining software?**

- **Is the workload developed and configured in a secure way?**

- **How are you monitoring security-related events in this workload?**

- **How is security validated and how do you handle incident response when a breach happens?**

- **How is connectivity secured for this workload?**

- **How have you secured the network of your workload?**

- **How are you managing encryption for this workload?**

- **Are keys, secrets and certificates managed in a secure way?**

- **What security controls do you have in place for access to Azure infrastructure?**

- **How are you managing identity for this workload?**

Once you answer the questions, as we have seen in the previous pillars, you will be presented with the overall review score:

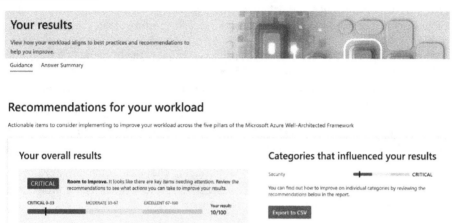

Figure 8.19 – Overall review score

The security recommendations will also be presented:

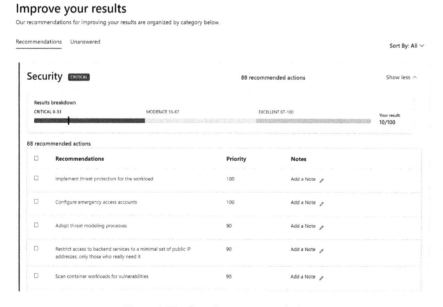

Figure 8.20 – Security recommendations

With this understanding of the questionnaire, let's now turn our attention to the reference architecture.

## Reference architecture

In our reference architecture, we have application tiers deployed to the same virtual network. Having said that, the web tier, business tier, and data tier have their own subnet, building isolation between tiers. The following recommendations can improve the security posture of our reference workload:

- Consider setting up an NVA to inspect the packets coming to the frontend of the application. Ideally, a landing zone will be set up and NVA will be hosted in the hub virtual network. Traffic will be routed to the virtual network hosting our application using **virtual network peering**.

- By default, traffic between resources within the same virtual network is allowed. However, this is not a desired state; either we can use an NSG to block traffic as required or use UDR to route traffic to NVA for inspection for all east-west traffic.

- Enable DDoS Protection to protect your public-facing endpoints and your workload from DDoS attacks. Azure virtual networks have basic DDoS protection; for better protection, it is better to upgrade to the Standard plan.

- Since the application needs to communicate with the database, there will be connection strings. Make sure you are using a key vault to store connection strings and secrets to connect to the database rather than storing keys in code.

- Enable Microsoft Defender for Servers to improve the security posture and to conduct vulnerability assessments of your workloads.

- In the reference architecture, we are using a jump box for managing servers in the virtual network. Since a VM is an IaaS solution, customers need to manage the security updates, patches, firewalls, and all other aspects of the operating system. It's better to upgrade to Azure Bastion, which is a PaaS solution letting you manage servers using RDP/SSH directly from a browser over a TLS connection.

- Enable the Web Application Firewall on Application Gateway to ensure that requests are inspected for web exploits and vulnerabilities.

- Leverage SQL encryption features such as transparent data encryption, column-level encryption, and backup encryption. Store the encryption keys in Azure Key Vault for better key management and automatic key rotation.

Voila! With that, we have successfully completed the security assessment of our reference workload. Also, this marks the end of this chapter and the book. Now, let's take a moment to summarize all the knowledge you've gained throughout this journey.

# Summary

In this last chapter, we have taken a systematic approach to assessing different pillars of the WAF. The assessment was conducted based on a questionnaire and a reference architecture. Throughout this chapter, we have delved into the depths of the WAF, examined the principles of each pillar, and contemplated the process of assessment. We started the chapter with an introduction to the assessment tool and reference architecture. The steps to be followed are common for all pillars when we are assessing them using the provided questionnaires, and later, specific steps are provided for each pillar. Each pillar has relevant questions and recommendations based on its reference architecture; depending on your environment, you need to apply the principles of each pillar.

As we turn the final pages of this book, let the principles of the WAF resonate in your mind and help you assess your workloads. May this book serve as a lens and compass to evaluate your workloads and implement the best practices outlined in the WAF. Remember, as new services and solutions emerge, new principles will be appended to the framework, so always make sure you follow the Microsoft documentation to stay updated. Nevertheless, feel free to tailor and personalize the recommendations based on your environment with the WAF principles in mind.

# Index

## A

# Z

Packtpub.com

Subscribe to our online digital library for full access to over 7,000 books and videos, as well as industry leading tools to help you plan your personal development and advance your career. For more information, please visit our website.

## Why subscribe?

- Spend less time learning and more time coding with practical eBooks and Videos from over 4,000 industry professionals

- Improve your learning with Skill Plans built especially for you

- Get a free eBook or video every month

- Fully searchable for easy access to vital information

- Copy and paste, print, and bookmark content

Did you know that Packt offers eBook versions of every book published, with PDF and ePub files available? You can upgrade to the eBook version at packtpub.com and as a print book customer, you are entitled to a discount on the eBook copy. Get in touch with us at customercare@packtpub.com for more details.

At www.packtpub.com, you can also read a collection of free technical articles, sign up for a range of free newsletters, and receive exclusive discounts and offers on Packt books and eBooks.

# Other Books You May Enjoy

If you enjoyed this book, you may be interested in these other books by Packt:

**FinOps Handbook for Microsoft Azure**

Maulik Soni

ISBN: 9781801810166

- Get the grip of all the activities of FinOps phases for Microsoft Azure
- Understand architectural patterns for interruptible workload on Spot VMs
- Optimize savings with Reservations, Savings Plans, Spot VMs
- Analyze waste with customizable pre-built workbooks
- Write an effective financial business case for savings
- Apply your learning to three real-world case studies
- Forecast cloud spend, set budgets, and track accurately

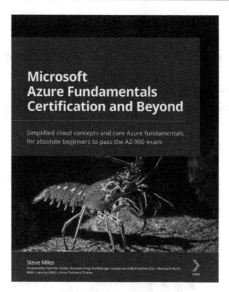

**Microsoft Azure Fundamentals Certification and Beyond**

Steve Miles

ISBN: 9781801073301

- Explore cloud computing with Azure cloud
- Gain an understanding of the core Azure architectural components
- Acquire knowledge of core services and management tools on Azure
- Get up and running with security concepts, security operations, and protection from threats
- Focus on identity, governance, privacy, and compliance features
- Understand Azure cost management, SLAs, and service life cycles

## Packt is searching for authors like you

If you're interested in becoming an author for Packt, please visit `authors.packtpub.com` and apply today. We have worked with thousands of developers and tech professionals, just like you, to help them share their insight with the global tech community. You can make a general application, apply for a specific hot topic that we are recruiting an author for, or submit your own idea.

## Share Your Thoughts

Now you've finished *Optimizing Microsoft Azure Workloads*, we'd love to hear your thoughts! Scan the QR code below to go straight to the Amazon review page for this book and share your feedback or leave a review on the site that you purchased it from.

https://packt.link/r/1837632928

Your review is important to us and the tech community and will help us make sure we're delivering excellent quality content.

# Download a free PDF copy of this book

Thanks for purchasing this book!

Do you like to read on the go but are unable to carry your print books everywhere?

Is your eBook purchase not compatible with the device of your choice?

Don't worry, now with every Packt book you get a DRM-free PDF version of that book at no cost.

Read anywhere, any place, on any device. Search, copy, and paste code from your favorite technical books directly into your application.

The perks don't stop there, you can get exclusive access to discounts, newsletters, and great free content in your inbox daily

Follow these simple steps to get the benefits:

1. Scan the QR code or visit the link below

https://packt.link/free-ebook/9781837632923

1. Submit your proof of purchase
2. That's it! We'll send your free PDF and other benefits to your email directly

www.ingramcontent.com/pod-product-compliance
Lightning Source LLC
Chambersburg PA
CBHW060545060326
40690CB00017B/3614